Researching Later Life and Ageing

Researching Later Life and Ageing

Expanding Qualitative Research Horizons

Edited by

Miranda Leontowitsch
St George's, University of London, UK

First published 2012 by
PALGRAVE MACMILLAN

Palgrave Macmillan in the UK is an imprint of Macmillan Publishers Limited, registered in England, company number 785998, of Houndmills, Basingstoke, Hampshire RG21 6XS.

Palgrave Macmillan in the US is a division of St Martin's Press LLC, 175 Fifth Avenue, New York, NY 10010.

Palgrave Macmillan is the global academic imprint of the above companies and has companies and representatives throughout the world.

Palgrave® and Macmillan® are registered trademarks in the United States, the United Kingdom, Europe and other countries

ISBN 978-0-230-28047-2

This book is printed on paper suitable for recycling and made from fully managed and sustained forest sources. Logging, pulping and manufacturing processes are expected to conform to the environmental regulations of the country of origin.

A catalogue record for this book is available from the British Library.

Library of Congress Cataloging-in-Publication Data
Researching later life and ageing : expanding qualitative research
    horizons / edited by Miranda Leontowitsch
      p. cm.
    ISBN 978–0–230–28047–2
    1. Aging–Research. I. Leontowitsch, Miranda, 1976–

    HQ1061.R4525 2012
    571.8′7872–dc23   2012023557

10 9 8 7 6 5 4 3 2 1
21 20 19 18 17 16 15 14 13 12

Printed and bound in Great Britain by
CPI Antony Rowe, Chippenham and Eastbourne

Contents

v

Acknowledgements

Although this volume is based solely on commissioned chapters without a preceding conference, the idea for this book was born at the Body, Ageing and Society Conference held in London in 2009.

I would like to thank Chris Gilleard, Paul Higgs, Iain Crinson and Lorna Ryan for their critical thoughts that helped shape the idea of the book, Jane Grogan for her encouragement and practical support in starting this project, Martin Hyde for reading my chapter and for the improvements I was able to make on the basis of his comments, and Grace Rose for her careful proof-reading of the entire manuscript and for providing the index. My thanks also go to the anonymous reviewer who provided valuable and encouraging comments on both the book proposal and manuscript. I am particularly grateful to the authors who have contributed to this volume and who responded to suggestions for revision in such a positive and prompt way.

Contributors

Christine Bigby is Director of Postgraduate Programmes in the School of Social Work and Social Policy at La Trobe University in Melbourne, Australia. She has a longstanding research interest in ageing and intellectual disability, and is the co-chair of the special interest ageing research group of the International Association for the Scientific Study of Intellectual Disability. She is editor of *Australian Social Work* and has published five books and numerous journal papers.

Laura Hurd Clarke is Associate Professor (Sociology) in the School of Kinesiology at the University of British Columbia, Vancouver, Canada. Dr Hurd Clarke has over a decade of research experience on the topic of the body and embodiment in later life, primarily in relation to ageism, appearance work, body image, health, and masculinity and femininity. In addition to a book entitled *Facing Age: Growing Older in an Anti-Aging Culture* (2010), she has published in such journals as *Ageing and Society, Canadian Journal on Aging, Journal of Aging Studies, Qualitative Health Research,* and *Sociology of Health and Illness.*

Paul Higgs is Professor of the Sociology of Ageing at University College London (UCL). He has published widely in both social gerontology and sociology. He is the co-author with Chris Gilleard of *Cultures of Ageing* (2000) and *Contexts of Ageing* (2005). In 2009 he published *Medical Sociology and Old Age* with Ian Rees Jones. He has researched topics including quality of life and consumption. Between 2009 and 2011 he coordinated an ESRC-funded seminar programme on 'New Ageing Populations' with Dr Karen Lowton (KCL) and Dr Karen Ballard (Surrey). He is currently researching the arena of intergenerational conflict. In addition Professor Higgs is joint editor of the journal *Social Theory and Health.*

Miranda Leontowitsch is a Lecturer in Qualitative Research Methods at St George's University of London, UK. Previously she worked at University College London after completing a PhD from Royal Holloway, University of London. She has researched the experiences of early retirement of senior executives, maintaining health in later life with over-the-counter and complementary medicines, and pharmacies as venues for health information. She is interested in critical and new approaches to using qualitative methods in researching later life and ageing, with a particular focus on researching older men.

Karen Lowton joined King's College London (KCL), UK in 2001 after completing a PhD at Royal Holloway, University of London focusing on how adults with cystic fibrosis and their family members perceive and manage the disease and its treatment. Karen has since studied the expectations and experiences of young people and health care staff concerning the transition to adult cystic fibrosis services, and how families of young adults who have died from cystic fibrosis have experienced end-of-life care and bereavement. Karen's research interests continue to focus on 'new' ageing populations, i.e. those reaching mid-life with traditional diseases or conditions of childhood, exploring health and social concerns, family care, and service use.

Mary MacMaster, retired Head Teacher of a first school, is writing up research for a PhD entitled 'Growing Old for Real: Women Image and Identity' at the Department of Fine Arts and Media Research at Norwich University College of the Arts (part thesis part photographic practice, see: www.growingold.co.uk). Her interest in photography developed over several years of City and Guilds evening class study before completing an MA in Photographic Studies in 2004 at Norwich University College of the Arts.

Sue Malta is a PhD candidate in the Faculty of Life and Social Sciences at Swinburne University of Technology in Melbourne, Australia. She is also managing editor of the online journal *International Journal of Emerging Technologies and Society* (iJETS) and recently edited the special edition on *Ageing and Technology* (Volume 8, No. 2, 2010: www.swin.edu.au/ijets). Sue's research interests include older adults, sexuality and the Internet, and slightly tangentially, social connectedness and social isolation. She is particularly interested in dispelling the idea that older adults are not sexual beings.

Barbara L. Marshall is Professor of Sociology at Trent University in Peterborough, Ontario, Canada where she teaches in the areas of sexuality, gender, the body, and social theory. She has written extensively on the medicalisation of sexuality, the pharmaceutical reconfiguration of sexual lifecourses, and the emergence of sexual functionality as an indicator of successful ageing. Her current research explores the ways gender and sexuality are embedded in accounts of ageing bodies across a range of different contexts, including sexual medicine, anti-ageing treatments, and public health promotion, and is opening up new questions about the sexualisation of the 'Third Age'.

Wendy Martin is a Lecturer in Health Studies at the School of Health Sciences and Social Care, Brunel University. Dr Wendy Martin is co-investigator on five funded research projects and Principal Investigator on the research project 'Photographing Everyday Life: Ageing, Lived Experiences, Time and Space' (ESRC First Grants scheme). She has recently completed, with Professor Christina Victor, an ESRC New Dynamics of Ageing Research Project on 'Families and Caring in South Asian Communities' and is involved with a Leverhulme-funded research project on inter- and intragenerational caring amongst minority communities within the United Kingdom.

Jane Seymour is a Sue Ryder Care Professor of Palliative and End-of-Life Studies, University of Nottingham, UK. She is Head of the Sue Ryder Care Centre for Palliative and End-of-Life Studies at the School of Nursing, Midwifery and Physiotherapy, University of Nottingham. Jane is a nurse and a social scientist, and has been involved in palliative care research and education since the early 1990s. She led (with Dr Katherine Froggatt, Lancaster University) a programme of research related to older people and end-of-life care within The Cancer Experiences Collaborative. Her research interests focus on advance care planning and other aspects of end-of-life decision-making, new models of palliative care, and public education in end-of-life care.

Christina Victor is Professor of Gerontology and Public Health in the School of Health Sciences and Director of the Health Ageing programme at the Brunel Institute for Ageing Studies (BIAS) at Brunel University. Her current research interests focus upon social relationships and later life and the experiences of ageing amongst ethnic minority elders. Christina has recently completed, with Dr Wendy Martin, an ESRC New Dynamics of Ageing Research Project on families and caring in South Asian communities and is currently working on a Leverhulme-funded project on inter- and intragenerational caring amongst minority communities.

Maria Zubair is a Research Associate at the School of Community Based Medicine at The University of Manchester. She has worked in an ESRC-funded New Dynamics of Ageing study 'Families and Caring in South Asian Communities' and is currently involved in a project investigating the under-use of health and social services by older South Asians with dementia. She is a member of the British Society of Gerontology and the Manchester Interdisciplinary Collaboration for Research on Ageing. Her areas of interest include race and ethnicity, gender, ageing, identity, social and familial relationships and support networks, and qualitative research methodologies.

Introduction

Miranda Leontowitsch

Research on ageing has predominantly relied on quantitative methods. This has been largely due to a political economy perspective that focused on poverty and ageing as a residual category. Thus, research was geared to measuring need and assessing ways in which health and social care could meet these in an economic way. Although the political economy focus has provided valuable insights into the plight of older people (and predominantly older women), it has led to viewing older people as a homogeneous group who live in deprived circumstances. The economic focus has been met by a biomedical one, which depicts ageing as a biological and inevitable downward trajectory of physical decline. Thus older people have been regarded as passive recipients of this economic and biological plight. However, later life has undergone considerable change over the past 40 years, including changes to employment, improvements in health and longevity, as well as fundamental changes to the social and cultural fabric of contemporary society (see Chapter 1 for a detailed account of these). This is not to suggest that poverty, social exclusion, inequalities and physical ageing in later life no longer exist, but it does call for research that takes these changed circumstances into account and acknowledges that older people are a highly heterogeneous group. The following quotation by Cook is an early recognition of what needed to change:

> If we want the public and the media to abandon the oversimplifying generalities they often make about age and aging and look instead at the diversity among older people, then gerontologists must stop asking attitudinal and factual questions about the elderly as if they were a homogenous group (Cook, 1992 in Thompson, 1994, p. 14).

Although there has been a steadily growing body of work in ageing that uses qualitative methods, the need for qualitative research continues. This is due to identifying more and more aspects of later life that have not been explored because they were thought of as irrelevant to older people. For example, in a review of studies of older people's participation in competitive sports, Dionigi (2006) points out that:

> the majority of research into this phenomenon has taken a quantitative approach or failed to consider older athletes' experiences in the context of broader sociocultural discourses. (…) The use of qualitative methods, such as in-depth interviews and observations, and interpretive analysis provided alternative ways of making sense of older adults and their relationships with competitive sport to what is typically found in the sport and aging literature (p. 365).

The paucity of qualitative research is also found in such areas as sexuality in later life, where the perception prevails that older people do not engage in sexual activity, or are reluctant to talk about intimate details of their lives (Gledhill & Abbey, 2008). Qualitative methods are particularly well equipped for mapping new research territories and for uncovering the more meaningful aspects of the lives of older people. With the methods available to qualitative researchers they can gain insights from the source most knowledgeable about later life, namely older people themselves. Phoenix and Smith (2011) examine how the 'master narratives' of passivity and decline are not necessarily matched by individual experiences of ageing.

> Counterstories are the stories which people tell and live that offer resistance to dominant cultural narratives. It is in their telling and living that people can become aware of new possibilities (M. Andrews, 2004). When told collectively, these 'new' stories present the possibility for both individual behavioural and social change (Phoenix & Smith, 2011, p. 630).

Thus, research on later life and ageing needs to continue its work on identifying which issues and aspects are important to older people, rather than relying on the top-down, quantifiable approaches that assume to know what constitutes later life.

The changes to later life have also been reflected in the composition of those considered to be old. For the first time those aged 65+ are no longer predominantly defined by women on a state pension or other forms of low income. As the post-World War II generation of men and women approach retirement age, the life expectancy gap between men and women narrows, and both groups enjoy increased longevity, more men live into old age. These changes are also reflected in the ethnic minority groups who have lived in the UK and many other European countries, and have decided to make these countries their place of retirement. At the same time, advances in biomedicine and technology have increased the life expectancy of many populations who historically did not survive childhood, such as those with cystic fibrosis. Although now living well into adult life, these new ageing populations remain unlikely to survive in good health to current pensionable age, and many marginalised groups such as older people with learning disabilities enjoy increased life expectancy but face the challenges of negotiating care and income with ageing carers and uncertain financial planning. All these voices, however, have been largely absent from ageing research. Within the next 20 years the proportion of older people from across these groups will significantly increase and shape our ageing society. Their experiences, concerns and needs are vital to understanding later life, today and in future. A new challenging question is how the experiences of these different groups of older people can be researched.

A review of the literature shows that qualitative methods are increasingly used in researching issues of later life, but that few authors reflect on their use of methods or provide a critical analysis of how qualitative methods (from sampling to data collection and analysis) need to be adapted in order to research a particular group of older people. This edited collection brings together authors from Australia, Canada and the United Kingdom, who provide a critical view of their own and current research practice. Moreover, they point to new research agendas, under-researched ageing populations, and old and new qualitative methods.

In Part I, Paul Higgs discusses the cultural and social processes that shaped the understanding of old age in the 20th century and how transitions in employment and social relations have substantially changed the lifecourse, thus the nature of later life and old age. This he argues, calls for research that goes beyond researching later life in terms of (biomedical) health and social policy, but also in terms of how older people experience and engage with their lives; lives that have become more complex due to changes in both the social relationships of later life and the societies in which they are experienced.

From this vantage point also, Laura Hurd Clarke draws attention to the continued disinterest and ambivalence towards the aged body in much gerontological research. Here too, the reluctance to consider the ageing body is rooted in ageist discourses and taken-for-granted assumptions about later life. With her extensive research experience she explores the challenges of asking both seemingly mundane questions about everyday activities associated with maintaining and dealing with the body, and deeply personal ones about sexuality, frailty and decline. In being reflective and open about how such research is and can be conducted, Laura Hurd Clarke encourages us to be role models for the next generation of researchers.

New forms of later life, and the centrality of the body provide the backdrop to Karen Lowton's chapter (Chapter 3) on new ageing populations. She maps the circumstances of new ageing populations and draws on her many interviews with people with cystic fibrosis and their relatives. Although managing a life-threatening chronic condition can at times be experienced as all-consuming, in-depth interviews provide insight into people's personal achievements in terms of health and social life, as well as into sensitive issues such as managing the likelihood of dying at a relatively early age.

The topic of new ageing populations introduces the second part of the book, in which research with older people from ethnic minority groups, and people with intellectual disability is discussed. Older men, who do not necessarily constitute a new group but who have been largely absent from research, are included here too. This part of the book is not solely concerned with raising attention to these under-researched groups, Chapters 4 and 5 in particular examine the challenges researchers face in gaining access to the field, dealing with gatekeepers, and gaining participants' trust. Moreover, Chapters 3, 4 and 5 challenge different prevailing assumptions about conducting qualitative research with members of marginalised groups. Maria Zubair, Wendy Martin and Christina Victor's chapter (Chapter 4) examines issues of researcher identity and the challenges of access and recruitment when researching older Pakistani Muslims in the UK. Zubair's experience as a young Pakistani Muslim woman researching older Pakistani men and women shows how an 'insider' position when researching co-ethnics cannot be assumed. By providing detailed insight and reflection on their fieldwork, the authors argue that an 'insider' relationship needs to be continuously and actively negotiated in the field through particular presentations of the embodied ethnic 'self'. With photographs of Maria Zubair in the way she dressed when conducting fieldwork, the authors explain why it was important for her to adopt a gendered Pakistani ethnic identity.

In line with Karen Lowton's chapter on new ageing populations, Christine Bigby maps the particular challenges people with intellectual disabilities face as they grow older, but also draws attention to the unexpected improvements some people with intellectual disability can experience in later life. She warns that the strong research focus on the carers (often parents) of older people with intellectual disability has served to give ownership of issues associated with ageing to parents rather than to the individuals themselves. With her longstanding interest in ageing and intellectual disability, Christine Bigby draws attention to locating hidden populations, getting past gatekeepers, using scaffolding techniques as a way of developing topic guides meaningful to people with an intellectual disability, as well as using participatory research with an onus on accessibility to research rather than training in research skills.

In Chapter 6, Miranda Leontowitsch examines why men have been largely absent from research, and the importance of masculinities in understanding older men's lives. She reviews the small literature on methodological issues in researching older men and then draws on her own experience of interviewing older men. She highlights the influence of gendered roles within qualitative interviewing, and how this enables her, as a younger female researcher, to gain rich data from a group of older men at the same time as protecting their sense of self.

Interviewing is by far the most popular method in qualitative research, and the same is true for qualitative studies in ageing research. Indeed Chapters 2 to 6 rely heavily on the interview method, showing how in-depth interviewing is a trusted way of learning about people's lives and experiences. However, this leaves a wealth of other qualitative methods largely ignored. The three chapters in the final part of this collection set out to illustrate how focus groups, online research and the use of photography can help illuminate the field of later life. Jane Seymour examines the advantages of using focus groups with older people in discussing end-of-life care issues. By focusing on four community studies she offers a reflective account of how issues associated with recruitment, informed consent and facilitating discussion of potentially distressing accounts in a group setting can be managed. She discusses the use of a vignette technique involving PowerPoint slides and its success. The chapter also includes excerpts from fieldnotes and quotations from focus group transcriptions that provide insight into the realities of conducting this kind of research.

In looking at new approaches to using qualitative methods, Sue Malta describes the process of conducting interviews with older people about new later life romantic relationships using instant messaging and email. She contributes to the slowly emerging literature on older peoples use of information and communication technology (ICT), and dispels the concern that older people are averse to being interviewed online. A computer screen shot, data excerpts, and detailed accounts of how the research was conducted provide ample material and thought for researchers embarking on this new method.

In the final chapter, Mary MacMaster explores ageing femininity through a combination of photographic self-portraits and staged images of women. Prior to taking the photographs she interviewed women about images of ageing and issues of identity. The self-portraits provide a window into understanding the experiences of older women in terms of identity and presentation of self, and the staged images combine Mary MacMaster's interpretation of what she heard and saw, with added symbolism by the women photographed through the use of personal objects.

Each chapter contains an annotated reading list and notes on titles the authors found helpful in conducting their research. Together with the range of research examples and ideas for practice, it is hoped that researchers in the field of later life will find themselves equipped with information and inspiration to help with their research, and the ability to continue the important reflexive work the authors have begun here.

References

Dionigi, R. (2006) 'Competitive sport and aging: The needs for qualitative sociological research', *Journal of Aging and Physical Activity*, 14 (4) 365–79.

Gledhill, S. & Abbey, J. (2008) 'Sampling methods: Methodological issues involved in the recruitment of older people into a study of sexuality', *Australian Journal of Advanced Nursing*, 26 (1) 84–94.

Thompson, E. (1994) 'Older men as invisible men in contemporary society' in E. Thompson (ed.) *Older Men's Lives*, pp. 1–21 (Thousand Oaks, CA: Sage).

Phoenix, C. & Smith, B. (2011) 'Telling a (good?) counterstory of aging: Natural bodybuilding meets the narrative of decline', *Journals of Gerontology, Series B*, 66 (5) 628–39.

Part I
Research Agendas

1
Later Life as an Arena of Change

Paul Higgs

This chapter situates contemporary later life in the social and cultural changes that have made the designation of old age problematic. In particular it argues that the social space in which older people now exist is substantially removed from the worlds of old age that existed for the majority of the 20th century in the industrialised world. The changes in employment and social relations which have transformed the most affluent countries in the world have also destandardised the lifecourse, making the distinction between adult life and old age unstable. This, in addition to growing affluence and improved health in retirement, has shifted the status of those in later life from occupying a residual category of health and social policy to a position more closely connected to mainstream cultural and societal processes. As significant in many ways as these structural changes has been the way in which cohort ageing has created a generational culture of later life namely the culture of the Third Age. This reconfiguration of old age moreover, has been facilitated by post-war babyboomers bringing their own dispositions and aspirations into retirement, and particularly by their not accepting the ascriptive status which previously circumscribed the old. Youth culture, leisure and the desire not to be defined as old has rewritten many of the scripts around old age, and thrown up new challenges for our understanding of this period of the lifecourse.

In this light, qualitative research has an important role in refocusing attention on later life as an important social and cultural space. In particular, attention to 'thicker descriptions' of the ways in which older people experience and engage with their lives provides many opportunities, not only to correct anachronistic assumptions about older people, but also to allow for more novel dimensions of this rapidly changing part of the lifecourse to be given the attention they deserve.

This chapter therefore is concerned with pinpointing the importance of studying later life through the cultural lens of qualitative research.

The changing nature of old age

It is commonplace to claim that we live in an ageing world. In Europe, Japan and North America it is very clear that larger numbers of people are living beyond the age of 65 and that, over the next few decades, this group is projected to increase both in size and as a proportion of the population. It has to be remembered, however, that this phenomenon is recent, reaching back to the second half of the 20th century at most. Before this, the numbers and proportions of older people in any given population were very much smaller, even if they exerted a disproportionate influence on the concerns of policymakers. Indeed it can be argued that the very creation of an ageing world – one where the overwhelming majority of people can expect to reach retirement age is a marker of the success of the modern world. It represents success on two counts: firstly, that there are now much lower rates of mortality in childhood and across the lifecourse (the so-called 'rectangularisation of the survival curve'); and secondly, because the institutionalisation of old age as retirement has made old age a chronologically specified point in people's lives. The ageing world is therefore a product of the modern world. The social status that might have once been accorded to the minority of older people in any community has now been transcended by status and policy issues surrounding retirement. It is in this context, of the social creation of old age through retirement, that we must situate contemporary ageing.

The standardisation of the lifecourse into socially organised periods emerged out of the transformations that accompanied industrialisation and the growth of capitalism. This was particularly noticeable in 19th century Europe and North America, where urbanisation and the industrial division of labour led to legislation regarding child labour, elementary education and more rigorous social policy. The intended and unintended effects of this was the creation of a standardised lifecourse in which childhood was more clearly separated from adulthood and seen as a period of education or preparation for the world of work. In a similar way the late 19th and early 20th centuries saw a more pronounced domestic division of labour, where men were increasingly expected to be 'breadwinners', and women to be 'homemakers'. As with childhood education, this position was to be buttressed by legislation as well as by social norms. This period of the late 19th and early 20th centuries also witnessed the emergence of the state retirement pension as a way of dealing with the effective redundancy of older workers in an environment dictated by the efficiency of labour power. As is well known the 'Iron Chancellor' Otto von Bismarck introduced the first state organised old age retirement pension in the 1870s in Germany, although it was not as popular as similarly introduced disability pensions. Other countries including New Zealand, Australia, Great Britain followed over the next few decades, so that by the 1930s most industrialised nations had a state retirement pension for men. Introduced for a variety of reasons these pensions had the impact of stabilising the position of old age within the societies in which they occurred, and over time (as the systems matured) took older people, particularly men, out of the labour market. In this way the idea of retirement as the final part of the standard lifecourse was established and became part of normative social structures.

What started out as a social policy response to the problem of the older worker has been transformed into a life stage of its own. Concern about social redundancy and poverty has been progressively shifted towards concern about the nature of this period of life. While the poverty of older people still continues to be an issue for many older people, this is often in the context of a general level of material prosperity. A marker of the shift in concerns can be seen in the way the position of the older person has been identified. The studies of Booth (1889) and Rowntree (1901) with their focus on the desperate economic circumstances of the old, can be contrasted with Talcott Parson's mid-20th century idea of the modern retiree's 'roleless role' (Parsons, 1942). Furthermore, much (particularly American) gerontological research from the 1950s started to focus on the determinants of 'successful ageing' (Palmore, 1979) rather than on its negative inevitability. Indeed the contrasting positions of 'activity theory' (Havinghurst & Albrecht, 1953) and 'disengagement theory' (Cumming & Henry, 1961) attest to the way that, within ageing research, the problem of being old was seen to be attitudinal rather than necessarily a result of poverty.

The 'selling of retirement' as a viable option for later life in the 1950s and 1960s in the United States of America (USA) was an outcome of both the emergence of Social Security pensions and private sector welfare policies (Hacker, 2002). In Britain it took longer, probably not until the 1980s, for the impact of occupational pension schemes to become realised (Hurd Clarke, 2006). However, what occurred in both countries was a transformation of the lives of large parts of the older population. Each successive cohort entered retirement with greater resources than had previous cohorts (Jones et al., 2008; Costa, 1998). Obviously there were inequalities and variations in this development, with non-white retirees in the United States of America (USA) faring much worse than their white counterparts, and women doing less well than men in terms of pension size. However, the key point was that instead of retirement being generally a period of 'structured dependency', it became a more diffuse entity. This process was much more marked in North America where more positive images of retirement jostled with more conventional depictions of old age. The existence of organisations such as the American Association of Retired People (now known as AARP), which focused as much on consumer discounts and entitlements as it did on social policy and health care, helped create the notion that retirement could be as much about living in Florida as living in poverty. As mentioned above, the desire to identify the causes of 'successful ageing' led to a whole series of research projects, including the Kansas City studies which investigated the dynamics of the emergent retired population (Neugarten & Associates, 1964). The history of these studies and their changing conclusions is less important than their recognition that the new experiences of retirement were ones that departed from constructs of the lifecourse dominant until this point. Retirement was not simply a residual category of life, rather, it was a period that needed to be understood in terms of its own ever-changing reality.

In a contrasting way, the idea that old age could be represented as structured dependency demonstrates the changes that had occurred to old age to that point. Coined by Peter Townsend the concept of structured dependency (Townsend, 1981) had its origins in the work of the social reformers Booth and Rowntree, and flowed from Townsend's own work on the social isolation of older people, and his broader interests in poverty (Townsend, 1957). An important dimension of his approach is the notion that the dependency of older people is 'structural' rather than solely an aspect of being old. In particular it is social policy that makes older people dependent and not the afflictions of age. Again, as with writers from North America, Townsend argued that the lives of retirees could be more socially engaged if it wasn't for the constraints that low state retirement pension entitlements place on older people. Noting the fact that the basic state pension did not keep pace with the rise in average earnings Townsend argues that it is unsurprising that the retired cannot participate in society to the degree they would like to. Interestingly however, the situation for retired people in the United Kingdom (UK) has improved greatly since the 1970s, with pensioner households seeing their incomes rise faster than those of the general population (Gilleard & Higgs, 2005). This paradox is a result of the fact that the income of the UK retired population has increased by 44 per cent in real terms between 1994/5 and 2008/9 higher than the growth in average earnings (ONS, 2011). In addition, large numbers of retirees in both the private and public sectors were benefiting from the maturity of occupational pension schemes which had begun in the 1950s and 1960s and which reflected more broadly increases in the standard of living (65 per cent of pensioner couples and 55 per cent of single pensioners received income from occupational pensions in 2008/9, ONS, 2011). Across the English Channel many welfare states chose not to implement what Gosta Esping-Andersen called the UK's 'flat rate universalism', but instead created more earnings-related public pension schemes, which meant that by the end of the 20th century the retired population were no longer seen as a social problem defined by their poverty, but rather now formed a social problem of intergenerational inequity (Esping-Andersen, 2002).

The Third Age

That the social world inhabited by the older person was changing was increasingly recognised by writers such as Peter Laslett (1989), who not only titled his contribution *A Fresh Map of Life*, but also popularised the idea of life after retirement as the 'Third Age'. This he saw as a period in which the combined benefits of increased healthy life expectancy (the compression of morbidity), and the greater resources available to retirees, led to this period being identified by him as 'the crown of life'. Laslett's work, for all its moralistic invocations of cultural guardianship is increasingly recognised as providing a conceptual location for the changes that were occurring to later life. The period after formal work and family responsibilities could now be seen as one organised around self-realisation and personal interests. Laslett's role in, and enthusiasm for, the now well established 'University of the Third Age' is merely one outcome of this re-orientation of focus on what constituted retirement. At other levels it provided a new nomenclature for older people themselves to use as alternatives to 'pensioner' and 'elderly'.

The idea of the Third Age has been very successful as a starting point for other writers seeking to re-position gerontology. Weiss and Bass (2002) have adapted the idea to provide a basis for what they term 'productive ageing'. As they write: 'Our impression is that many in the Third Age find life to be enjoyable and satisfactory yet often feel themselves not fully engaged' (Weiss & Bass, 2002, p. 5). Here the security of the retired population is set against a sense of lacking social worth. The echoes of Laslett's views are very noticeable, and this line of thinking has been utilised to counter some of the generational burden arguments raised in the USA.

However, a more sociologically nuanced interpretation of the Third Age has been made by Gilleard and Higgs (2005, 2011a) who transformed the Third Age from being regarded as a moral exhortation or a code for the activities of well off older people (the tritely named 'woopies'). Instead, Gilleard and Higgs see the Third Age operating as a cultural space in which contemporary ageing occurs. Not only has there been change in the mortality and morbidity rates of older populations, but the improved standard of living and exposure to consumer culture has changed how people experience being older. A critical factor is also the nature of the cohorts who are now entering retirement at the present moment. For Gilleard and Higgs these are better understood as providing the basis for a generational culture, a culture formed in the key decade of the 1960s, and which is changing the social terrain as it ages. Moving beyond the simple formulation of age, period and cohort effects Gilleard and Higgs introduce the notion of 'generational habitus', amalgamating the ideas of Karl Mannheim and Pierre Bourdieu on generation and habitus respectively. The generational habitus created by the youth culture of the 1960s was not only a powerful example of the growth of consumerism in terms of fashion, music and goods, but also rested on a 'generational schism' regarding the desirability of social stability. The dispositions and cultural milieu surrounding generational habitus emphasised choice, freedom and lifestyle over the more structured and constrained virtues of post-war society. That these dispositions fitted well with the growth of a consumer society is often overlooked in accounts of the importance of the 1960s in transforming social relations. It has also to be accepted that for many, the changes that this period brought about were not immediately evident. Their impact gradually made its way through society, to the extent that the need to have an identity as a consumer in contemporary society is evidenced from the choice of supermarket to the activities of both local and national governments, where the 'freedom' to have a 'choice' is deemed to be a paramount virtue.

If an engagement with consumer society is one of the outcomes of this transformative period it is also the case, according to Gilleard and Higgs, that the Third Age is constituted by the generational field of those sixties cohorts. The ideas of choice and freedom as well as a broadening of youth culture have meant that retirement can be the site of a variety of lifestyle opportunities – so much so that notions of early retirement start to figure as a desirable choice and not an illness-related termination of employment. Identifying the Third Age as a cultural field connected to the genersational habitus of post-1960s cohorts helps explain some of the cultural tropes that have recently come to represent the collapsing boundaries of old age. A clear example of this is popular music, which no longer has either a shelf life connected to the period in which it was popular or to the age of those first exposed to it. A feature of the 'generational schism' was the need for the young to break from the cultural patterns of their parents. This was all the more effective if it also shocked them. Such distinctions, let alone an emphasis on rejecting the past, are not the objectives of either the youth or the older generation of today. Rather there is both a desire to break down boundaries and evidence of an infinite recycling of 1960s music and culture, to all born after that decade. In part this testifies to the ageing of youth culture itself with its emblematic image of youth rebellion, the Rolling Stones now all being of pensionable age. But it is also the case that much more is evidenced, namely that we now inhabit a much more diffuse cultural world where old age is often culturally effaced or only articulated in terms of the profound dependencies of frailty and death.

In their account Gilleard and Higgs draw on the ideas of 'reflexive modernisation' articulated by Beck and his colleagues (Beck et al., 2003) into a research programme seeking to investigate the emergence of what they term a 'second modernity'. Beck's ideas about the transformation of the social world through a set of 'revolutions by side effects' has many implications for the study of contemporary later life. Central to the idea that we live in a period of second modernity is that many of the issues we face are ones that have been created by rational responses to earlier difficulties created by industrialisation and capitalism. The welfare state and policies established social security and entitlement schemes, in which certain issues were seen from particular perspectives. Hence, pensions were seen in terms of poverty reduction and labour inefficiency. As we have seen, these initial formulations were superceded in the main by the creation of retirement as a stage of life, potentially understood in terms of the culture of the Third Age.

Researching the changing nature of ageing

Within this formulation the study of later life also goes through a similar change in focus. For many decades research on ageing has focused on the narrative of decline associated with the effects of the 'ageing process'. Moving beyond the boundaries of biomedical research considerable effort has been expended on charting the conditions and circumstances of older people. The community or social survey has played a key role in providing social policy with information about the older population as a group, and has sparked both reform and reorganisation of services. This approach does suffer from one major drawback which makes it less useful in contemporary circumstances; older age is identified through the prism of health and social policy. In other words, the circumstances of old age are reduced to need for social care, pensions or health care interventions. Again, as was pointed out above, the emergence of old age as a separate social category owes much to the discourses set up by social reformers and health professionals. While many of the outcomes of these initiatives have been positive, the contexts of ageing in which later life is now lived have been correspondingly transformed, and so naturally must the topics of research investigation.

Consumption

For example, the study of the consumption patterns of older households has had relatively little attention paid to it unless undertaken for the purposes of estimating household poverty. In the UK this has meant that profound and far reaching improvements to older peoples' standard of living have been left unresearched. Research utilising government data has shown that retired households participated in the consumption revolution that occurred from the 1980s, and were as likely to own important consumer goods such as colour televisions and video cassette recorders and take holidays abroad, in a broadly similar fashion to younger cohorts (Higgs et al., 2009). All this information was available but rarely utilised for these purposes. Instead, up until the end of the 20th century it was assumed that most older people remained cut off from participation in consumer society. Detailed analysis of consumption and ownership data not only allowed for a more detailed picture of the position of older people to emerge but it also allowed for a true reckoning of the role played by both period and cohort in shaping these patterns. In a similar fashion data captured by the English Longitudinal Study of Ageing (ELSA) has been able to address the more complicated question of the role of information and communication technology (ICT) in making older people feel closer or more detached from their neighbourhoods (Gilleard & Higgs, 2008). Again the role of ICT was not one simply of creating a 'digital divide' between older and younger populations; rather, use of mobile phones and the internet could be a way that older people could be more included. Such complex relationships are important to the study of how later life is currently experienced (see Chapter 8), but such research questions do not emerge spontaneously out of the 'ageing as a product of social policy' approach.

Indeed, it appears that much of the research on ageing positively adopts a moral tone in relation to what is identified as the 'frivolous' aspects of contemporary ageing, namely, those aspects that do not seem to connect with the agendas of health, income or social value. Again, taking an example from research into the consumption practices of retired people, gerontologists seem averse to examining the 'serious leisure' involved in taking holidays. This is all the more remarkable given that tourism researchers have identified the 50+ age group as one of the key components of the global tourist industry. It is unsurprising that therefore, in many countries, organisations such as SAGA and the AARP (who provide a range of services and products for this demographic) are probably more well known to older people than the age focused charities or campaigning groups. Again, in the UK many 'ageing as a product of social policy' approach researchers have consequently minimised the importance of retirement migration and seasonal migration to key locations such as the Mediterranean (King et al., 2000). A few key individuals have drawn attention to this aspect of later life, mainly because they have moved away from the narrower 'methodological nationalism' that besets this approach. It is indeed ironic that the significance of those British people who have chosen to retire or winter in countries such as Spain only emerges in relation to age-related universal benefits such as the UK's Winter Fuel Allowance, which is paid to all UK pensioners whether they live in Scotland or the Costa Brava. Researchers in both the USA and Australia have been more responsive to the issue of 'Snowbirds' and 'Grey Nomads' for considerably longer than has been the case in the UK, and it could be conjectured that this attention might result from a greater awareness of the diversity of circumstances in retirement in these countries (Katz, 2005; Higgs & Quirk, 2007).

Quantitative and qualitative research

The effect of the policy orientated paradigm of old age has meant a fascination with the quantification of need through various forms of measurement. This has created a burgeoning research industry as the rationalisation of health and social care policy has demanded its own indices and standardised responses, and has downplayed other forms of research, particularly qualitative ones. Quality of Life (QoL) research has been a particular example of this, as study after study has attempted to find non-health-related (or reducible) elements of this elusive dimension. It is therefore not surprising that there is an industry of researchers producing scales which can be applied to different groups of older people, located in different circumstances, and with varying degrees of dependency and/or illness. Paradoxically, there is a role for qualitative research within this paradigm, but it is limited, as is too often the case, to the piloting of the dimensions of the putative scale. For QoL scales to be regarded as being properly grounded, there is often a requirement that the components that make up the instruments, are derived from the views and attitudes of the subject population themselves, and this needs to be gleaned from the 'thicker' description provided by qualitative research. While much biomedical and health-related research is content to restrict itself to this limited role for non-quantitative approaches, there is gradually an awareness that, as in other areas of social life, there is a richness to older people's lives that not only repays attention, but also tells us much more about the complexities of modern societies.

Taking as a first example underdeveloped research in the area of older people's engagement with discourses of health, we can see that many studies of the most profound chronic conditions affecting those in later life tend to exclude or ignore the older population (Higgs & Jones, 2009). While Twigg (2006) has addressed this issue and while some qualitative health researchers have looked at issues such as arthritis (Sanders et al., 2002), in the main age has not been a factor to be considered, other than in the context of decline. Even work examining the discourses of the 'will to health', which have become popular over the last decade seem to see this as an age-limited engagement. Yet all the evidence suggests that with the erosion of assumptions about the 'natural' lifecourse have come changes in the way people are encouraged to view their health in later life (Jones & Higgs, 2010). Added to this must be the profound 'pharma-ceuticalisation' of much health care, a considerable proportion of which is aimed at older people or at the difficulties associated with growing older. Within gerontology there have been studies looking at the nar-ratives surrounding anti-ageing medicine, and there have been some studies examining the nature of the engagement with this (Slevin, 2010). Other researchers have addressed just how the older woman has to negotiate the physical signs of ageing, whether this be through modes of dressing or the colouring of hair (Hurd Clarke et al., 2009; Hurd Clarke & Korot-chenko, 2010). Sometimes this research is negative towards the whole enterprise and sometimes it is aware of the contradictory nature, but more often than not it is absent from the agenda of health researchers.

An area where there has been some engagement with the themes of embodiment and ageing is in the area of how health interacts with phys-ical exercise and sport. Connecting positive health promotion messages regarding the benefits of exercise and involvement in sporting activities, such as running, with broader issues surrounding the negotiation of ageing identities, researchers like Tulle (2008), Phoenix and Smith (2011), and Phoenix et al. (2005) have interrogated narratives about the role of the ageing body in contextualising engagement with various health practices. For some, participation in sport can be a way of coming to terms with age, while for others it is a form of resisting the ascriptive status of old age. That the discourses are complex and often contradictory makes further work in this area both interesting and essential if full cognisance of the changing nature of old age in contemporary society is to be achieved. This complexity extends to the emerging field of the neuro-cultures of ageing, and in particular to the fear of developing cognitive problems such as dementia. Not only does this area throw up some of the issues already raised about anti-ageing techniques but it also has connections with the fear of loss of agency, a fear that has led to new opportunities for the marketing of pharmaceutical products to offset conditions such as Mild Cognitive Impairment (MCI) (Williams et al., 2012).

Increasingly these debates return to a question that is at the heart of the new reflexive conditions which is, 'what is normal ageing?'. Jones and Higgs (2010) have argued that not only has there been a gradual separation between the 'natural' lifecourse, or indeed cycle, and the 'normal' experience of ageing, a separation which has problematised many normative assumptions about old age, but that also the structures of normativity have broken down. Following Ulrich Beck (2007) they argue that instead of social institutions creating the conditions for the reproduction of normative values, social institutions have become subject to the diversity of social life and reproduce the notion that it is diversity itself that is normative. The example that Beck gives of this process is the change from the nuclear family as the established norm affirmed by social institutions from welfare, education, employment and law to the current situation, where in most countries, a variety of family household structures are validated and deemed equal. Obviously not all options are treated equally, with many individuals and communities holding on to a more 'modernist' set of attitudes about the acceptability of this diversity of choice, but Beck's point is that the notion of diversity of choice is now projected at an institutional level, with all its concomitant implications for 'choosing wisely'.

If the normativity of diversity is now an underpinning of the institutional arrangements of a reflexive second modernity, then such discourses also affect ageing and old age. As the normal in 'normal' ageing becomes more diverse it is not surprising that there are as many narratives of ageing as there are cultures of ageing. The 'normativity of old age' on which policies such as state funded or supported Old Age Pensions are based become seen as anachronistic at best, and unfair at worst. Moves to increase the age of eligibility for these benefits have been successfully implemented in many countries, and changes to the financial arrangements of retirement are being constantly discussed by policymakers hoping to 'offset' long-term commitments. How older people negotiate these new circumstances is now presented as an expression of choice as well as the exercise of freedom. The culture of the Third Age is part of this reflexivity, but is also enmeshed in the pursuit of forms of 'successful ageing' which, while accepting of the various lifestyle choices making up later life, also views them through the lens of *not* getting old. Jones and Higgs (2010) therefore see that the various practices that make up the 'will to health' in later life are ones that validate a continuing engagement with agency, rather than an acquiescing to passivity and dependency. A pursuit of an unobtainable 'fitness' at many different levels is a linking motif for the Third Age, whether that is in terms of clothing, physical appearance or indeed in terms of sexual potency. This is not to deny that many are not in a position to engage with these discourses, or indeed that some may choose to reject them as undesirable objectives. This inability or refusal to engage is not necessarily a problem. In the wider context of the culture of the Third Age such choices may seem 'flawed', but again, they are reincorporated as part of the diversity of normativity now surrounding ageing.[1] This notion of choice and diversity does not seemingly stop at the very boundaries of life and death. The growing demand that people should be free to choose not only their end of life care, but also in some cases, whether or not they are free to plan to end their own lives at a pre-ordained time and place.

   The agenda of choice and agency is not one that has been conventionally associated with later life, rather, it is often seen as being connected to younger people whose lives are not yet so firmly embedded in work and family life. Consequently, when topics such as these are approached by social gerontologists they are often seen as only being relevant within a discourse of providing dignity to dependent older people within particular settings such as nursing (Gilleard & Higgs, 2010). What the change in both the social relationships of later life as well as the societies in which they are experienced has achieved, is to create more complex circumstances that need researching in ways that give full voice to their complexity and diversity.

Conclusion

What this chapter has attempted to do is to provide an overview of how the circumstances and experience of old age have been transformed over the past century. It has described a situation where old age moved first from the margins of society to a position of relative security (albeit one seen as only concerning health and social policy) and subsequently became not only a generalised expectation for the vast majority, but through the Third Age, one of its key sites of cultural engagement. In all of this, old age transformed into later life, mirroring many of the changes that have been occurring at a more general societal level as life becomes more concerned with identity, consumption and lifestyle. As a consequence, it is not surprising that ageing, embodiment and culture are important concerns in studying later life. As concerns of simply surviving in old age have shifted to concerns of how to positively experience later life, then the subject matter of gerontology, as well as its research methods, need to change. These transformations not only require researchers to adapt to engaging with this new environment, but also to examine the assumptions they bring to an emerging and changing research area.

Notes

1 The absence of any discussion of those whose lack of agency not only deems them dependent but in fact the concern of health and social care practitioners. This is not to see them as unimportant but rather it is to recognise that they are affected by a different discourse – that of the Fourth Age. See Gilleard and Higgs (2010, 2011b) for a discussion of the implications of ageing without agency.

Annotated reading list

Laslett, P. (1989) *A Fresh Map of Life* (London: Weidenfield and Nicholson).

This is a key reference in discussing the changing nature of modern later life and introducing the terms Third Age and Fourth Age. Laslett seeks to demonstrate that the long Third Age is both a demographic and personal achievement. His idea is very much located in a moral understanding of what the duties of older people are and this has resonated with writers seeking to develop the 'productive ageing' approach.

Townsend, P. (1981) 'The structured dependency of the elderly: A creation of social policy in the 20th century', *Ageing and Society*, 1, 5–28.

This is one of the foundational texts of the 'structured dependency' approach to ageing which shares many assumptions with what has come to be called 'critical gerontology'. The paper focuses on the role of social policy in creating the negative experiences of older people and in particular points out that dependency is structured and not just a function of growing older.

Gilleard, C. & Higgs, P. (2011) 'The third age as a cultural field' in D. Carr & K. Komp (eds) *Gerontology in the Era of the Third Age: Implications and Next Steps*, pp. 33–50 (New York: Springer).

This chapter is a condensation of arguments made by Gilleard and Higgs in a number of books and articles since the publication of *Cultures of Ageing* in 2000. Specifically it seeks to make the study of the Third Age more sociologically grounded by relating it less to personal aspiration and more to generational habitus. In particular it stresses the importance of consumption and embodiment in making up the cultural field of the Third Age.

Katz, S. (2005) *Cultural Ageing* (Peterborough, Ontario: Broadview).

This collection of essays by Stephen Katz provides an overview of how the contemporary experience of ageing has changed and how it offers a number of different ways of studying it. Chapters in the book cover areas as diverse as Canadian retirement communities in Florida; the use of drugs to combat erectile dysfunction; and the management of activity in everyday life.

References

Beck, U. (2007) 'Beyond class and nation: Reframing social inequalities in a globalizing world', *British Journal of Sociology*, 58 (4) 679–705.

Beck, U., Bonss, W. & Lau, C. (2003) 'The theory of reflexive modernization: Problematic, hypotheses and research programme', *Theory, Culture & Society*, 20 (2) 1–33.

Booth, C. (1889) *Life and Labour of the People* (London: Macmillan).

Costa, D. (1998) *The Evolution of Retirement* (Chicago: University of Chicago Press).

Cumming, E. & Henry, W.E. (1961) *Growing Old* (New York: Basic Books).

Esping-Andersen, G. (2002) 'Towards the good society, once again' in G. Esping-Andersen (ed.) *Why We Need a New Welfare State*, pp. 1–25 (Oxford: Oxford University Press).

Gilleard, C. & Higgs, P. (2010) 'Theorizing the fourth age: Aging without agency', *Aging and Mental Health*, 14, 121–8.

Gilleard, C. & Higgs, P. (2008) 'Internet use and the digital divide in the English longitudinal study of ageing', *European Journal of Ageing*, 5, 233–9.

Gilleard, C. & Higgs, P. (2005) *Contexts of Ageing: Class, Cohort and Community* (Cambridge: Polity).

Gilleard, C. & Higgs, P. (2011a) 'The third age as a cultural field' in D. Carr & K. Komp (eds) *Gerontology in the Era of the Third Age: Implications and Next Steps*, pp. 33–50 (New York: Springer).

Gilleard C. & Higgs P. (2011b) 'Aging, abjection and embodiment in the fourth age', *Journal of Aging Studies*, 25 (2) 135–42.

Hacker, J. (2002) *The Divided Welfare State: The Battle over Public and Private Social Benefits in the United States* (Cambridge: Cambridge University Press).

Havinghurst, R. & Albrecht, R. (1953) *Older People* (New York: Longmans).

Higgs, P., Hyde, M., Gilleard, C., Victor, C., Wiggins, R. & Jones, I.R. (2009) 'From passive to active consumers? Later life consumption in the UK from 1968–2005', *Sociological Review*, 57, 102–24.

Higgs, P. & Jones, I.R. (2009) *Medical Sociology and Old Age* (London: Routledge).

Higgs, P. & Quirk, F. (2007) '"Grey nomads" in Australia: Are they a good model for successful ageing and health', *Annals of the New York Academy of Sciences*, 1114, 251–7.

Hurd Clarke, G. (2006) 'The UK occupational pension scheme in crisis' in H. Pemberton, P. Thane & N. Whiteside (eds) *Britain's Pensions Crisis: History and Policy* (London: British Academy).

Hurd Clarke, L., Griffin, M. & Maliha, K. (2009) 'Bat wings, bunions, and turkey wattles: Body transgressions and older women's strategic clothing choices', *Ageing and Society*, 29 (5) 709–26.

Hurd Clarke, L. & Korotchenko, A. (2010) 'Shades of grey: To dye or not to dye one's hair in later life', *Ageing and Society*, 30 (6) 1011–26.

Jones, I.R. & Higgs, P. (2010) 'The natural, the normal and the normative: Contested terrains in ageing and old age', *Social Science and Medicine*, 71 (8) 1513–19.

Jones, I.R., Hyde, M., Victor, C., Wiggins, D., Gilleard, C. & Higgs, P. (2008) *Ageing in a Consumer Society: From Passive to Active Consumption in Britain* (Bristol: Policy Press).

Katz, S. (2005) *Cultural Ageing* (Peterborough, Ontario: Broadview).

King, R., Warnes, A. & Williams, A. (2000) *Sunset Lives: British Retirement Migration to the Mediterranean* (Oxford: Berg).

Laslett, P. (1989) *A Fresh Map of Life* (London: Weidenfield and Nicholson).

Neugarten, B. & Associates (1964) *Personality in Middle and Late Life: Empirical Studies* (New York: Atherton).

ONS (Office of National Statistics) (2011) *Pensions Trends 2011* (London: ONS).

Palmore, E. (1979) 'The predictors of successful aging', *Gerontologist*, 19, 427–41.

Parsons, T. (1942) 'Age and sex in the social structure of the United States', *American Sociological Review*, 7, 604–16.

Phoenix, C. & Smith, B. (2011) 'Telling a (good) counterstory of aging: Natural bodybuilding meets the narrative of decline', *Journals of Gerontology B*, 66 (5) 628–38.

Phoenix, C., Faulkner, G. & Sparkes, A.C. (2005) 'Athletic identity and self-ageing: The dilemma of exclusivity', *Psychology of Sport and Exercise*, 6, 335–47.

Rowntree, B.S. (1901) *Poverty: A Study in Town Life* (London: Macmillan).

Sanders, C., Donovan, J. & Dieppe, P. (2002) 'The significance and consequences of having painful and disabled joints in older age: Co-existing accounts of normal and disrupted biographies', *Sociology of Health and Illness*, 24, 227–53.

Slevin, K.F. (2010) '"If I had lots of money…I'd have a body makeover": Managing the aging body', *Social Forces*, 88 (3) 1003–20.

Townsend, P. (1957) *The Family Life of Old People* (London: Routledge, Kegan Paul).

Townsend, P. (1981) 'The structured dependency of the elderly: A creation of social policy in the 20th century', *Ageing and Society*, 1, 5–28.

Tulle, E. (2008) 'Acting your age? Sports science and the ageing body', *Journal of Aging Studies*, 22, 340–7.

Twigg, J. (2006) *The Body in Health and Social Care* (Basingstoke: Palgrave).

Weiss, R.S. & Bass, S.A. (eds) (2002) *Challenges of the Third Age: Meaning and Purpose in Later Life* (New York: Oxford University Press).

Williams, S., Higgs, P. & Katz, S. (2012) 'Neuroculture, active ageing and the "older brain": Problems, promises and prospects', *Sociology of Health and Illness*, 34, 64–78.

2
Researching the Body and Embodiment in Later Life

Laura Hurd Clarke

In 1993, Chris Shilling sagely pointed out that the body had 'historically been something of an "absent presence"' (p. 15) in the social sciences. Since that time, there has been 'a veritable explosion of interest' (Williams & Bendelow, 1998, p. 1) in the body and embodiment within socio-cultural theorising and research. However, the subjective experience of having and being in an ageing and aged body has not been as extensively examined (Arber & Ginn, 1991; Calasanti & Slevin, 2001; Faircloth, 2003; Hurd Clarke, 2010; Tulle, 2008; Twigg, 2004). Additionally, there has been limited elucidation of the process and complex nature of conducting research on older adults' perceptions and experiences of their ageing bodies. In this chapter, I examine the reasons for the continued disinterest in and ambivalence towards the aged body and refer to key examples of research that have positioned the older body as an essential focal point in gerontology and the sociology of ageing. I will also discuss the challenges of investigating the doing of gender and age (West & Zimmerman, 1987) in and through the body, drawing on my own experiences of conducting research with men and women about their bodies and embodied experiences in later life.

The coming of the ageing body in qualitative research

If one surveys the past scope of research in social gerontology, it quickly becomes apparent that the body has largely and curiously been invisible, while simultaneously being indirectly present and undeniably important. While scholars have probed the nature of ageism in the workplace, the use of body stereotypes as a source of humour in birthday cards, the burden of caregiving for the elderly, various aspects of health and illness in later life, or older adults' interactions with health care professionals, the body has often been something of a proverbial elephant in the academic room – neither acknowledged nor truly seen. Even the, albeit limited in number though not in importance, methodological works focused on the qualitative study of ageing (see, for example, Gubrium & Sankar, 1994; Reinharz & Rowles, 1988) have not attended to the complexities of researching older adults' perceptions of and experiences in their ageing bodies.

There are several theories for why the body has been a relative late-comer to socio-cultural studies of ageing. Twigg (2004) asserts that gerontologists have often deliberately avoided the body in their efforts to reframe ageing as a social rather than physiological process. According to this line of thinking, scholarly attention to the body was considered to be a 'retrogressive step, one that [took] us back into the territory of biological determinism and the narrative of decline' (Twigg, 2004, p. 60). Twigg (2004) contends that gerontologists were concerned that attention to the ageing and aged body would demean older adults, by reducing them to mere bodily characteristics, thereby often reinforcing 'the misery perspective' (Oberg, 2003, p. 103). Rather than buttressing stereotypes that depicted later life as a time of inevitable and all-encompassing physical loss and functional dependence, or defining older people solely in terms of their corporeal realities, socio-cultural scholars shifted their research focus to other aspects of ageing, such as the social structural inequalities that mitigated the experience of growing older.

Additionally, the academic silence surrounding ageing and aged bodies have derived from internalised ageism, or what Twigg (2004) aptly refers to as 'gerontophobia' (p. 60). Thus, some researchers and theorists (see, for example, Calasanti, 2005; Calasanti & Slevin, 2001; Oberg, 2003; Twigg, 2004) have asserted that the avoidance of studying the ageing body stems from the (perhaps unconscious) acceptance of ageist portrayals of later life and the concomitant societal ambivalence towards the physical realities of growing older. As well as the personal implications of internalised ageism, existing cultural discomfort with aged corporeality generally and, more specifically, with physical decline, loss of attractiveness, sexual devaluation and increasing dependence, has until recently translated into a scholarly uneasiness with, if not avoidance of, the ageing body as a focus of research (Arber & Ginn, 1991).

Whether the failure to attend to the aged body has arisen from the desire to resist ageist discourses or from the internalisation of those same discriminatory and oppressive societal norms, it is clear that gerontologists lag behind other disciplines in theorising and researching corporeality. Nevertheless, there is a growing and exciting vein of research within qualitative gerontology that has begun to mine the experience of having an ageing body within anti-ageing western culture. This relatively new sub-field within social gerontology has been 'part of a wider Cultural Turn' (Twigg, 2004, p. 60) in which social constructionism has yielded important insights into how the ageing body is 'formed and given meaning within culture' (p. 60). Leading the way have been theorists such as Cruikshank (2003), Gilleard and Higgs (2000), Gullette (1997, 2004) and Katz (1996). At the same time, theorising about the socially constituted nature of the ageing body has been accompanied by a growing number of qualitative studies focused on three broad categories of scholarship: the embodied experience of illness in later life; the everyday experience and management of the aged body; and older adults' perceptions of their appearances.

To begin, there is an extensive and rich tradition of research that examines embodiment in relation to various health issues among older adults, originating with Bury's (1982) landmark and much-cited study of the experience of rheumatoid arthritis and the concept of 'biographical disruption' (Bury, 1982, p. 167). The ensuing cadre of research has explored how the body is taken for granted or outside personal awareness until physical declines alter the person's daily life and sense of embodied identity. Highlighting the interactions between the body, notions of masculinity and femininity, identity and health, the research in this area has examined the experience of conditions typically associated with old age, including: arthritis (Bury, 1988; Gibbs, 2008; Rosenfeld & Faircloth, 2004; Sanders et al., 2002); Alzheimer's disease and dementia (Kontos, 2004; Phinney & Chesla, 2003); heart disease (Husser & Roberto, 2009); osteoporosis (Roberto, 1990; Wilkins, 2001); Parkinson's disease (Gisquet, 2008; Solimeo, 2008; Stanley-Hermanns & Engebretson, 2010); prostate cancer (Chapple & Ziebland, 2002; Oliffe & Thorne, 2007); and the effects of stroke (Becker, 1993; Becker & Kaufman, 1995; Faircloth et al., 2004).

The second area of corporeal research has centred on the management and everyday experience of the ageing body (see, for example, Faircloth, 2003). This line of study is best epitomised by Twigg's (1997, 2000) nuanced examination of the washing and bathing of predominantly old bodies in the context of community care programmes. Using data from interviews with home care workers and recipients of home care residing in London and a coastal town in the United Kingdom, Twigg's (2000) study highlighted how the seemingly mundane management of the body is embedded in gender, class, age and race discourses, as well as in power inequalities. Twigg's (2000) study has been followed by research concerned with topics like food and eating (Moss et al., 2007); sleep (Hislop

& Arber, 2003, 2006); oral hygiene and health (McKenzie-Green et al., 2009); exercise, physical activity and sport participation (Bundon et al., 2011; Krekula, 2007; Poole, 2001; Tulle, 2003, 2008; Wainwright & Turner, 2003); and sexuality (Gott & Hinchliff, 2003; Hinchliff & Gott, 2008; Hurd Clarke, 2006; Krekula, 2007; Potts et al., 2006; Rosenfeld, 2003; Slevin, 2006).

Finally, Furman's (1997) poignant ethnography of Julie's International Salon in the United States signaled the first extensive foray into understanding the meanings that older women attributed to their appearances and self-presentation strategies. Based on 18 months of ethnographic fieldwork, including participant observation and informal interviewing, her book provided rich insights into beauty shop culture and the many ways that her female participants were concerned about and responded to 'the loss of youth, slenderness, wrinkle-free faces and natural hair colour' (p. 173). Furman (1997) revealed how the women's perceptions and experiences of their bodies were often shaped and constrained by the privileging of youthfulness, to the detriment of their self-esteem, cultural currency and even social visibility. Furman's study was followed by studies primarily concerned with older women's perceptions of and feelings about their ageing appearances (see, for example, Dumas et al., 2005; Fairhurst, 1998; Hurd, 1999, 2000; Hurd Clarke, 2010; Krekula, 2007; Paulson & Willig, 2008; Slevin, 2006; Winterich, 2007) and their associated beauty work efforts and attitudes (Hurd Clarke & Bundon, 2009; Hurd Clarke & Griffin, 2007, 2008a; Hurd Clarke & Korotchenko, 2010; Hurd Clarke et al., 2007; Slevin, 2010; Ward & Holland, 2011). While there have been some studies exploring how older men perceive their ageing bodies (Baker & Gringart, 2009; Hurd Clarke et al., 2008; Kaminski & Hayslip, 2006; Oberg & Tornstam, 1999), this literature is noticeably limited and most of the research has not focused exclusively on men's experiences. Indeed, research concerned with the embodied meanings that older men attribute to their changing bodies, including appearance and functional abilities, is woefully lacking.

In summary, research focused on the subjective experience of ageing and having an aged body is a much-needed new sub-field in social gerontology and the sociology of the body. This broad and rich field of study has illuminated the ambivalence with which aged bodies are viewed and experienced as well as the complexities of having an older body in a youth, health and image focused society. Nevertheless, there is room for further theoretical development, nuanced investigation and methodological and academic innovation, in order to more fully explicate the lived realities of having and being in an aged, gendered body.

The unique challenges of researching the ageing body

Gubrium and Holstein (2003) have contended that the body is not a static compilation of body parts, but is rather 'an unfinished experiential entity' (p. 205) that is continually interpreted and reinterpreted in social interactions. In this way, the body is subjective, situated in everyday life and 'a material entity suffused with meaning' (Gubrium & Holstein, 2003, p. 207). However, Twigg (2000) notes that many aspects of corporeality are difficult to research because they 'exist at a level that is rarely brought into conscious articulation or review; indeed in modern western societies we are largely educated to ignore them, regarding them as too trivial or too private for comment' (p. 4). Twigg (2000) cogently argues that it is imperative that we explore the mundane body and its management 'if we are to grasp something of the day-to-day textures of people's lives and the sources of implicit meaning and significance that are embedded in them' (p. 4). Despite the emergent scholarship concerning ageing and aged bodies, little has been written about the process of conducting research in the area. In this section, I will draw on the existing methodological offerings that have been articulated by various scholars working in the area of ageing and the body, as well as proffering some of my own insights and experiences related to three methodological issues, namely: 1) the aged body as a sensitive subject; 2) internalised ageism; and 3) the emotional risks and costs of doing this type of research.

The aged body as a sensitive subject

Given societal discomfort with and disparagement of ageing and aged bodies, it is unsurprising that research on the subject invariably engenders ambivalence on the part of the researcher and the researched (Hurd Clarke, 2003). Indeed, uneasiness about, or even revulsion towards the aged body and the accompanying lines of inquiry, can make it difficult to establish rapport and feel comfortable asking or answering interview questions. While backstage (Goffman, 1959) accounts of the research process are incorporated into the methods sections of graduate theses and dissertations and frequently constitute the fodder of question and answer periods at academic conferences, there are few published accounts of how to negotiate the invariably sensitive nature of researching the ageing body. One laudable exception can be found in the work of Furman (1997) who reported that 'asking women to reflect on their facial wrinkles and other marks of ageing was too intrusive and intimidating a request' (p. 10) as it 'got too close to the vulnerabilities women experience as they age' (p. 10). To address this issue, Furman employed photo elicitation, whereby she had the women select current photos of themselves as well as pictures from their youth and middle age. Rather than directly querying the women about how they felt about their changing physical appearances, she asked the women to describe their life histories as they were depicted in the photographs as well as to reflect on their evolving looks. Furman (1997) stated that 'by treating the photograph as a kind of artefact, participants were able to gain some distance from it and to feel less self-conscious' (p. 10). Certainly, visual methods such as Photovoice (Wang, 1999) afford qualitative researchers additional means of acquiring insight and facilitating discussion, if not social change, about sensitive issues and experiences of embodied inequality and oppression (see Chapter 9).

Similarly, other strategies for establishing rapport and facilitating the discussion of difficult or taboo topics include reassuring participants about confidentiality at the outset of the interview (Kaufman, 1994; Legard et al., 2003) and beginning each interview session with easily answered questions (Britten, 1995; Rubin & Rubin, 2005). To the latter end, I always begin my initial interviews by asking participants to describe their life histories, recounting those details they feel are important and necessary for providing a context for the ensuing conversation (Hurd Clarke, 2003). While I have found that some peoples' stories are fraught with tragedy and sadness (Hurd Clarke & Griffin, 2008b) and therefore, that this topic can be unexpectedly emotionally taxing, for the most part the men and women I have spoken with have found it relatively easy, if not pleasurable, to describe the biographical particulars of their lives, often with great relish and in extensive detail. Moreover, the provision of an historical and evolutionary context for the individuals' current embodied experiences has served the vital importance of elucidating the complex and situational meanings that participants attribute to various aspects of their bodies.

Additionally, Legard et al. (2003) aptly challenge researchers to be aware of whose problem the purportedly contentious topic actually is as they state:

> Any unease or embarrassment on the part of the researcher will communicate itself to the participant and may make them reticent about discussing the topic…Researchers will often be surprised at how willing people are to talk about sensitive subjects and how their own discomfort seems to be greater than that of the interviewee (p. 162).

Undoubtedly, I have found this to be true in my own interviews focused on body image and sexuality, a topic I will return to in the next section.

However, even when participants appear at ease while discussing sensitive topics, I strongly encourage all researchers not to conclude interviews in potentially emotionally charged terrain, which might cause individuals to feel overly exposed or regretful of their candour. Hey (1999) points out that if participants are left in an emotionally vulnerable place or with the sense that they have been too revealing of their intimate thoughts and feelings, the researcher's departure and the conclusion of the interview 'may well mean consigning elderly people back to a heightened aware-ness of their social isolation' (p. 107). Thus, towards the end of the inter-view I strive to gently move from delicate topics such as feelings about one's body, the experience of loss and frailty, sexuality, or death and dying, into more cerebral discussions about the issues at hand, such as asking participants to reflect on the future implications of a particular topic or policy related to the scope of the interview. In this way, I endeav-our to end the interview on a more philosophical note, which simultan-eously moves the discussion into a broader social context. Additionally, Alty and Rodham (1998) sagely suggest that interviews on sensitive topics should conclude with the researcher asking participants if there are any remaining issues or topics they might wish or need to discuss, as well as offering to provide additional resources, such as counselors or informal support groups, where appropriate. To that end, researchers should be well informed of their community resources prior to interviews.

Internalised ageism

Inherently related to the definition and experience of sensitive topics, internalised ageism invariably influences both the researcher and the researched to some (not always conscious) degree. While this dynamic may result in taken-for-granted assumptions and stereotypes about later life embodiment, internalised ageism can inadvertently shape the questions we ask and the way we frame our lines of inquiry. A moment I often reflect on as a turning point in my own thinking about ageism and the conduct of research, came after my first interview for my doc-toral study (concerned with women's body image and ageing) with an 82-year-old woman. Specifically, she telephoned me several days after the interview and had this to say:

> I was talking with some friends about your study and our meeting on Monday. They wanted to know what you asked me about. Anyway, well, we got hysterical at this one friend who asked me several times, 'What about a question in regard to sexual interests?' And you know, I really was surprised that you didn't ask me about sex. Sex is an

important part of how you feel about your body. So I want to know – in the second interview are we going to talk about sex? Because I've got something to say!

Admittedly, I was taken aback and embarrassed by the comment, because it was the last thing I expected and, more importantly, I realised that the fact I had not asked about sexuality was deeply revealing of me and my own internalised ageism. Although I had asked around the issue of sexuality, by questioning the woman about her relationship with her deceased husband, I had been hesitant and unsure of how to bring up what I had erroneously assumed would be a sensitive topic, a taboo subject, or an irrelevant issue for a widowed, aged woman. My ageist assumptions were all the more egregious given my purported research focus and expertise. The woman's directness and openness were acutely humbling and instructive and from then on, I made a point of asking the other 21 women in the study about their experiences and perceptions of sexuality. In addition to stories of passion and sexual satisfaction, I was privy to accounts of sexual assault, betrayal, physical discomfort during intercourse, incompatible libidos and sexual yearnings. While the majority of the women indicated that they were comfortable and willing to discuss their sexual experiences and perceptions, several women expressed discomfort. For example, the last woman I interviewed declined to discuss the topic because she felt it was 'a private issue' and one that did not influence her own experience of her ageing body. Nevertheless, I concluded it was better to ask questions thoughtfully and respectfully and allow people the option to decline to discuss sex and sexuality, rather than assume automatically that sex and sexuality were non-existent, if not non-issues.

So how does one remain aware of internalised ageism as an underlying factor shaping the research experience? Certainly, we are not always privileged to have participants such as the woman I described above, who keep us on track and challenge us to confront our own inherent and often concealed bigotry or misguided assumptions. Nevertheless, self-awareness, constant personal reflection through strategies such as journaling, continually reading the work of established and new scholars, communication with both participants and research team members and the willingness to receive and grow from constructive feedback from others are imperative. As soon as we as researchers begin to assume that we have our substantive topic, our method and our theoretical underpinnings completely figured out and in order, we will invariably have reached a stage of unacceptable complacency.

The emotional toll of research

While it is also a privilege, if not a sacred trust, there is no denying how emotionally taxing it can be to listen to or simply bear witness to another human being's suffering or sense of loss, whether that be in relation to illness experiences, the everyday management of the body, or changing appearances and resultant feelings of low self-esteem. The literature is full of examples of how distressing it can be to research sensitive topics (Campbell, 2002; Dickson-Swift et al., 2006, 2007; Higgins, 1998; Lee, 1993). Lee (1993) simply puts it this way: 'The interview is typically a stressful experience for the interviewee and the interviewer' (p. 102). A particularly poignant example of the emotional costs of doing research on later life embodiment can be found in the writing of Higgins (1998), who described her experiences as a researcher in a nursing home, where she faced the bleak realities of institutional life and was attendant as people described raw pain and often conveyed their intense and unrelenting physical and emotional misery. Higgins (1998) noted how demanding, if not overwhelming, her work as a researcher was:

> The sights I see and the stories the elderly people tell me, shake the foundations of my being, of my past; it brings back many forgotten experiences. During the day, I cry with [the participants] as they cry. At night, I cry for all elderly people as I confront the prospect of my own ageing mother and mother-in-law and what the future might mean for them. I relive the deaths of my grandparents and my lonely father. I am confronted by my own ageing and mortality. For the elderly people in my study, I feel an overwhelming sense of pity, helplessness and powerlessness (p. 864).

Higgins' (1998) account of her research experiences reveals how painfully difficult it is to interview individuals who are frail, dependent, lonely and dispossessed, particularly as we all potentially face a similar future. However, Higgins' (1998) forthrightness is relatively uncommon, as we typically refrain from expressing these types of academic, if not emotional, accounts and they are infrequently published in scholarly venues.

My own experiences of being a researcher have been very similar at times, as I have heard countless deeply moving stories about debilitating health issues, emotional abuse, sexual and physical assault, incalculable personal losses, self-loathing, isolation, loneliness, personal regrets and fears about and wishes for death and dying. However, I have rarely voluntarily shared my personal reflections on what it is like to hear these stories with a public audience through conference presentations or scholarly publications, unless I attend qualitative methods conferences, as I have wanted to privilege what my participants have had to say rather than appearing to be self-indulgent. Sensitive topics, especially those underscored by internalised ageism, have been ones that have challenged my sense of equilibrium as a researcher and as a woman now in her forties. For example, I have heard many women express profound dislike of their bodies and the physical realities of growing older, such as one 73- year-old woman who had this to say:

> I hate my hands. I hate my feet. I hate my belly. I hate my bum. I hate my boobs. So, now there's nothing – I don't really like anything. I used to like it because I thought I was fairly tall…but now, I don't think there's any part of my body that I like…I always look in the mirror and think I look horrible…I don't like my body.

The power of ageist and sexist discourses on the above woman's perceptions of her body, let alone her daily existence, were shocking and dismaying to me, as I simultaneously felt her anguish and was forced to reflect on what the future might hold as my own appearance increasingly moved away from idealised feminine beauty.

Undoubtedly, it is extremely important for researchers studying ageing embodiment to find ways to be aware of and cope with their own resultant feelings in productive and healthy ways that simultaneously protect the confidentiality and dignity of the participants and enable themselves to engage in the process analytically, yet compassionately. In other words, as Hey (1999) notes, the process of conducting the research entails 'the challenges of listening to harrowing stories and yet being able to "detach" and "walk away"' (p. 105). Most of the sparse literature that addresses how to cope with the emotional costs of doing such research advocates various means of debriefing. For example, Higgins (1998) turned to her dissertation supervisor and her husband for support, whereas Davis (2001) used her field notes and her university seminars as a meaningful outlet for her distress at witnessing suffering and death in a hospital setting. I, myself, have benefitted from supportive mentors and have, in turn, listened (with what I hope has been compassion and thoughtful guidance) to numerous graduate students and some peers as they, often tearfully, recounted their own difficult research experiences and painful feelings about the same. In addition to writing field notes about my research encounters and using physical activity as a form of personal release and meditative practice, I have also employed a form of journaling and have encouraged my students to do the same. Specifically, I have used my research journal to identify and process my own emotions about my interactions with study participants, the stories they share with me and my impressions of their personal situations. Not only has my research journal allowed me to find an appropriate place to reflect on and express my personal feelings, but it has also served as an important part of my research analysis. Indeed, Davis (2001) has pointed out that her own emotions, as well as those of her participants, further clarified her understanding of the experiences of health care workers in their daily work lives, as well as permitting emotions to become a topic of research in itself. In this way, our emotions are a product of our research encounters, as well as an important part of the data and the co-creation of findings with our participants.

Nevertheless, Johnson and Clarke's (2003) interviews with scholars engaged in research concerned with sensitive topics revealed that some individuals experience feelings of isolation and a lack of support as they struggle to manage their own emotions, debrief from their difficult research experiences and separate their work lives from their private lives. Johnson and Clarke's (2003) study highlights the fact that we can and should do more as a research community to help each other identify and work through the ways in which the act of doing research affects our lives in both positive and negative ways. Clearly, we need to create safe places, where scholars at all stages of their careers can appropriately share and process, not only their findings, but also the myriad of ways that they are personally touched and changed by the act of doing research.

Concluding comments

From the mundane aspects of daily embodied life, to the resiliencies and miseries associated with being in an unhealthy or frail older body, to attitudes about ageing appearances, research on aged corporeality is still in its infancy. Building on the extant literature, future research is needed to more fully explore the nuanced meanings that older adults and those in their immediate lives attribute to their bodies and the management of the same. For instance, topics such as oral care, the possession and employ of dentures, glasses and other assistive devices, toileting and incontinence, the use and experience of clothing, eating, perceptions and responses to appearance change and the relationships between ageism and various embodied experiences warrant further investigation. Additionally, the research on the ageing body and embodiment needs to more fully analyse the impact of intersectionality and the diversity of experiences across gender, culture, age, sexuality and social class.

At the same time, researchers need to convey more about their experiences in the field and the strategies they employ for overcoming the difficulties they encounter. Research on the ageing body is both exciting and underscored by ambivalence and internalised ageism, which make the process of investigation tricky, personally challenging and often emotional. While I am not inviting extreme forms of navel gazing, it would behoove the research community to follow the excellent lead of scholars like Furman (1997) and Twigg (2000) and be more forthcoming about the process of doing research on ageing and the body. By conveying the creative and sometimes serendipitous solutions we devise for dealing with the complicated and frustrating travails of establishing rapport, facing our own demons and managing our identities and our emotions, we can collectively move the field forward, open ourselves to the contributions of others and act as role models for future generations of researchers.

Lest my comments unwittingly scare off potential scholarly recruits, it is important that I conclude by speaking to the intrinsic rewards of engaging in research on ageing and the body. As well as having the honour of listening to other peoples' life and corporeal stories and benefiting from their acquired wisdom, experience and willingness to share, I have found my chosen field of research to be intellectually fascinating, personally growth motivating and professionally gratifying as I have met many wonderful and inspiring people along the way. That said, as anti-ageing discourses continue to take further hold in western culture and later life becomes ever more devalued and feared, the study of the ageing body will become more fraught with tensions as well as personal, academic and societal discomfort, sensitivities and ambivalence. Nevertheless, the study of the ageing body and aged embodiment has the potential to deepen our understanding of what it means to grow old, as well as to explore the depth and insidious nature of ageist discourses.

Annotated further reading

Dickson-Swift, V., James, E.L., Kippen, S. & Liamputtong, P. (2007) 'Doing sensitive research: What challenges do researchers face?', *Qualitative Research*, 7, 327–53.

This novel study utilises data from in-depth interviews with 30 researchers to explore their perceptions of the challenges associated with the qualitative investigation of sensitive topics.

Furman, F.K. (1997) *Facing the Mirror: Older Women and Beauty Shop Culture* (New York: Routledge).

This wonderful ethnography of beauty shop culture is replete with rich qualitative data and reflexive commentary on the process of doing research on ageing, the body and embodiment.

Gilbert, K.R. (ed.) (2001) *The Emotional Nature of Qualitative Research* (New York: CRC Press).

This excellent edited collection explores the emotionally laden nature of scholarly investigation, providing insights into how researchers manage, utilise and benefit from this often unreported part of the research encounter.

Twigg, J. (2000) *Bathing – The Body and Community Care* (London: Routledge).

Written by one of the leading scholars in the sociology of aging and the body, this beautifully crafted book illuminates the centrality of the body in community care and affords the readers numerous and important methodological insights.

References

Alty, A. & Rodham, K. (1998) 'The ouch! factor: Problems in conducting sensitive research', *Qualitative Health Research*, 8, 275–82.

Arber, S. & Ginn, J. (1991) *Gender and Later Life: A Sociological Analysis of Resources and Constraints* (London: Sage).

Baker, L. & Gringart, E. (2009) 'Body image and self-esteem in older adulthood', *Ageing and Society*, 29, 977–95.

Becker, G. (1993) 'Creating continuity after a stroke: Implications of life course disruption', *The Gerontologist*, 33, 148–58.

Becker, G. & Kaufman, S.R. (1995) 'Managing an uncertain illness trajectory in old age: Patients' and physicians' views of stroke', *Medical Anthropology Quarterly*, 9, 165–87.

Britten, N. (1995) 'Qualitative interviews in medical research', *British Medical Journal*, 311, 251–3.

Bundon, A., Hurd Clarke, L. & Miller, W.C. (2011) 'Frail older adults and patterns of exercise engagement: Understanding exercise behaviours as a means of maintaining continuity', *Qualitative Research in Sport, Exercise and Health*, 3, 33–47.

Bury, M. (1982) 'Chronic illness as biographical disruption', *Sociology of Health and Illness*, 4, 167–82.

Bury, M. (1988) 'Meanings at risk: The experience of arthritis' in R. Anderson & M. Bury (eds) *Living with Chronic Illness: The Experience of Patients and Their Families* (London: Unwin Hyman).

Calasanti, T. (2005) 'Ageism, gravity and gender: Experiences of aging bodies', *Generations*, 29, 8–12.

Calasanti, T.M. & Slevin, K.F. (2001) *Gender, Social Inequalities and Aging* (New York: Altamira).

Campbell, R. (2002) *Emotionally Involved: The Impact of Researching Rape* (New York: Routledge).

Chapple, A. & Ziebland, S. (2002) 'Prostate cancer: Embodied experience and perceptions of masculinity', *Sociology of Health and Illness*, 24, 820–41.

Cruikshank, M. (2003) *Learning to be Old: Gender, Culture and Aging* (Lanham, MD: Rowman and Littlefield).

Davis, H. (2001) 'The management of self: Practical and emotional implications of ethnographic work in a public hospital setting' in K.R. Gilbert (ed.) *The Emotional Nature of Qualitative Research* (New York: CRC Press).

Dickson-Swift, V., James, E.L., Kippen, S. & Liamputtong, P. (2006) 'Blurring boundaries in qualitative health research on sensitive topics', *Qualitative Health Research*, 16, 853–71.

Dickson-Swift, V., James, E.L., Kippen, S. & Liamputtong, P. (2007) 'Doing sensitive research: What challenges do researchers face?', *Qualitative Research*, 7, 327–53.

Dumas, A., Laberge, S. & Straka, S.M. (2005) 'Older women's relations to bodily appearance: The embodiment of social and biological conditions of existence', *Ageing and Society*, 25, 883–902.

Faircloth, C.A. (2003) *Aging Bodies: Images of Everyday Experience* (Walnut Creek, CA: Altamira).

Faircloth, C.A., Boylstein, C., Rittman, M., Young, M.E. & Gubrium, J. (2004) 'Sudden illness and biographical flow in narratives of stroke recovery', *Sociology of Health and Illness*, 26, 242–61.

Fairhurst, E. (1998) '"Growing old gracefully" as opposed to "mutton dressed as lamb": The social construction of recognizing older women' in S. Nettleton & J. Watson (eds) *The Body in Everyday Life* (New York: Routledge).

Furman, F.K. (1997) *Facing the Mirror: Older Women and Beauty Shop Culture* (New York: Routledge).

Gibbs, L. (2008) 'Men and chronic arthritis: Does age make men more likely to use self-management services?', *Generations*, 32, 78–81.

Gilleard, C. & Higgs, P. (2000) *Culture of Ageing: Self, Citizen and the Body* (London: Prentice-Hall).

Gisquet, E. (2008) 'Cerebral implants and Parkinson's disease: A unique form of biographical disruption?', *Social Science and Medicine*, 67, 1847–51.

Goffman, E. (1959) *The Presentation of Self in Everyday Life* (Garden City, New York: Doubleday and Company).

Gott, M. & Hinchliff, S. (2003) '"How important is sex in later life?" The views of older people', *Social Science and Medicine*, 56, 1617–28.

Gubrium, J.F. & Holstein, J.A. (2003) 'The everyday visibility of the aging body' in C.A. Faircloth (ed.) *Aging Bodies: Images and Everyday Experience* (New York: Altamira).

Gubrium, J.F. & Sankar, A. (eds) (1994) *Qualitative Methods in Aging Research* (Thousand Oaks, CA: Sage).

Gullette, M.M. (1997) *Declining to Decline: Cultural Combat and the Politics of Midlife* (Charlottesville, VA: The University Press of Virginia).

Gullette, M.M. (2004) *Aged by Culture* (Chicago: The University of Chicago Press).

Hey, V. (1999) 'Frail elderly people: Difficult questions and awkward answers' in S. Hood, B. Mayall & S. Oliver (eds) *Critical Issues in Social Research: Power and Prejudice* (Philadelphia: Open University Press).

Higgins, I. (1998) 'Reflections on conducting qualitative research with elderly people', *Qualitative Health Research*, 8, 858–66.

Hinchliff, S. & Gott, M. (2008) 'Challenging social myths and stereotypes of women and aging: Heterosexual women talk about sex', *Journal of Women and Aging*, 20, 65–81.

Hislop, J. & Arber, S. (2003) 'Understanding women's sleep management: Beyond medicalization-healthicization?, *Sociology of Health and Illness*, 25, 815–37.

Hislop, J. & Arber, S. (2006) 'Sleep, gender and aging: Temporal perspectives in the mid-to-later life transition' in T.M. Calasanti & K.F. Slevin (eds) *Age Matters: Realigning Feminist Thinking* (New York: Routledge).

Hurd, L. (1999) '"We're not old!": Older women's negotiation of aging and oldness', *Journal of Aging Studies*, 13, 419–39.

Hurd, L. (2000) 'Older women's body Image and embodied experience: An exploration', *Journal of Women and Aging*, 12, 77–97.

Hurd Clarke, L. (2003) 'Overcoming ambivalence: The challenges of exploring socially charged issues', *Qualitative Health Research*, 13, 718–35.

Hurd Clarke, L. (2006) 'Older women and sexuality: Experiences in marital relationships across the life course', *Canadian Journal on Aging*, 25, 129–40.

Hurd Clarke, L. (2010) *Facing Age: Women Growing Older in Anti-Aging Culture* (Lanham, MD: Rowman and Littlefield).

Hurd Clarke, L. & Bundon, A. (2009) 'From "the thing to do" to "defying the ravages of age": Older women reflect on the use of lipstick', *Journal of Women and Aging*, 21, 198–212.

Hurd Clarke, L. & Griffin, M. (2007) 'The body natural and the body unnatural: Beauty work and aging', *Journal of Aging Studies*, 21, 187–201.

Hurd Clarke, L. & Griffin, M. (2008a) 'Body image and aging: Older women and the embodiment of trauma', *Women Studies International Forum*, 31, 200–8.

Hurd Clarke, L. & Griffin, M. (2008b) 'Visible and invisible ageing: Beauty work as a response to ageism', *Ageing and Society*, 28, 653–74.

Hurd Clarke, L., Griffin, M. & the PACC Research Team (2008) 'Failing bodies: Body image and multiple chronic conditions in later life', *Qualitative Health Research*, 18, 1084–95.

Hurd Clarke, L. & Korotchenko, A. (2010) 'Shades of grey: To dye or not to dye one's hair in later life', *Ageing and Society*, 30, 1011–26.

Hurd Clarke, L., Repta, R. & Griffin, M. (2007) 'Non-surgical cosmetic procedures: Older women's perceptions and experiences', *Journal of Women and Aging*, 19, 69–87.

Husser, E.K. & Roberto, K.A. (2009) 'Older women with cardiovascular disease: Perceptions of initial experiences and long-term influences on daily life', *Journal of Women and Aging*, 21, 3–18.

Johnson, B. & Clarke, J.M. (2003) 'Collecting sensitive data: The impact on researchers', *Qualitative Health Research*, 13, 421–34.

Kaminski, P.L. & Hayslip, B. (2006) 'Gender differences in body esteem among older adults', *Journal of Women and Aging*, 18, 19–35.

Katz, S. (1996) *Disciplining Old Age: The Formation of Gerontological Knowledge* (Charlottesville, VA: The University Press of Virginia).

Kaufman, S.R. (1994) 'In-depth interviewing' in J.F. Gubrium & A. Sankar (eds) *Qualitative Methods in Aging Research* (Thousand Oaks, CA: Sage).

Kontos, P.C. (2004) 'Ethnographic reflections on selfhood, embodiment and Alzheimer's disease', *Ageing and Society*, 24, 829–49.

Krekula, C. (2007) 'The intersection of age and gender: Reworking gender theory and social gerontology', *Current Sociology*, 55, 155–71.

Lee, R.M. (1993) *Doing Research on Sensitive Topics* (Newbury Park, CA: Sage).

Legard, R., Keegan, J. & Ward, K. (2003) 'In-depth interviewing' in J. Ritchie & J. Lewis (eds) *Qualitative Research Practice: A Guide for Social Science Students and Researchers* (Thousand Oaks, CA: Sage).

McKenzie-Green, B., Giddings, L.S., Buttle, L. & Tahana, K. (2009) 'Older peoples' perceptions of oral health: "It's just not that simple"', *International Journal of Dental Hygiene*, 7, 31–8.

Moss, S.Z., Moss, M.S., Kilbride, J.E. & Rubinstein, R.L. (2007) 'Frail men's perspectives on food and eating', *Journal of Aging Studies*, 21, 314–24.

Oberg, P. (2003) 'Image versus experience of the aging body' in C. Faircloth (ed.) *Aging Bodies: Images and Everyday Experiences* (Walnut Creek: Altamira Press).

Oberg, P. & Tornstam, L. (1999) 'Body images among men and women of different ages', *Ageing and Society*, 19, 629–44.

Oliffe, J. & Thorne, S. (2007) 'Men, masculinities and prostate cancer: Australian and Canadian patient perspectives of communication with male physicians', *Qualitative Health Research*, 17, 149–61.

Paulson, S. & Willig, C. (2008) 'Older women and everyday talk about the ageing body', *Journal of Health Psychology*, 13, 106–20.

Phinney, A. & Chesla, C.A. (2003) 'The lived body in dementia', *Journal of Aging Studies*, 17, 283–99.

Poole, M. (2001) 'Fit for life: Older women's commitment to exercise', *Journal of Aging and Physical Activity*, 9, 300–12.

Potts, A., Grace, V.M., Vares, T. & Gavey, N. (2006) '"Sex for life"? Men's Counterstories on "erectile dysfunction", male sexuality and ageing', *Sociology of Health and Illness*, 28, 306–29.

Reinharz, S. & Rowles, G.D. (eds) (1988) *Qualitative Gerontology* (New York: Springer Publishing Company).

Roberto, K.A. (1990) 'Adjusting to chronic disease: The osteoporotic woman', *Journal of Women and Aging*, 2, 33–47.

Rosenfeld, D. (2003) 'The homosexual body in lesbian and gay elders' narratives' in C.A. Faircloth (ed.) *Aging Bodies: Images of Everyday Experience* (Walnut Creek, CA: Altamira).

Rosenfeld, D. & Faircloth, C. (2004) 'Embodied fluidity and the commitment to movement: Constructing the moral self through arthritis narratives', *Symbolic Interaction*, 27, 507–29.

Rubin, H.J. & Rubin, I.S. (2005) *Qualitative Interviewing: The Art of Hearing Data*, 2nd edn (Thousand Oaks, CA: Sage).

Sanders, C., Donovan, J. & Dieppe, P. (2002) 'The significance and consequences of having painful and disabled joints in older age: Co-existing accounts of normal and disrupted biographies', *Sociology of Health and Illness*, 24, 227–53.

Shilling, C. (1993) *The Body and Social Theory* (Newbury Park, CA: Sage).

Slevin, K.F. (2006) 'The embodied experiences of old lesbians' in T.M. Calasanti & K.F. Slevin (eds) *Age Matters: Realigning Feminist Thinking* (New York: Routledge).

Slevin, K.F. (2010) '"If I had lots of money...I'd have a body makeover": Managing the aging body', *Social Forces*, 88, 1003–20.

Solimeo, S. (2008) 'Sex and gender in older adults' experience of Parkinson's disease', *Journal of Gerontology: Social Sciences*, 63B, S42–48.

Stanley-Hermanns, M. & Engebretson, J. (2010) 'Sailing the stormy seas: The illness experience of persons with Parkinson's disease', *The Qualitative Report*, 15, 340–69.

Tulle, E. (2003) 'The bodies of veteran elite runners' in C.A. Faircloth (ed.) *Aging Bodies: Images of Everyday Experience* (Walnut Creek, CA: Altamira).

Tulle, E. (2008) *Ageing, the Body and Social Change: Running in Later Life* (London: Palgrave Macmillan).

Twigg, J. (1997) 'Deconstructing the "social" bath: Help with bathing at home for older and disabled people', *Journal of Social Policy*, 26, 211–32.

Twigg, J. (2000) *Bathing – The Body and Community Care* (London: Routledge).

Twigg, J. (2004) 'The body, gender and age: Feminist insights in social gerontology', *Journal of Aging Studies*, 18, 59–73.

Wainwright, S.P. & Turner, B.S. (2003) 'Aging and the dancing body' in C.A. Faircloth (ed.) *Aging Bodies: Images of Everyday Experience* (Walnut Creek, CA: Altamira).

Wang, C. (1999) 'Photovoice: A participatory action research strategy applied to women's health', *Journal of Women's Health*, 8, 185–92.

Ward, R. & Holland, C. (2011) '"If I look old, I will be treated old": Hair and later-life image dilemmas', *Ageing and Society*, 31, 288–307.

West, C. & Zimmerman, D.H. (1987) 'Doing gender', *Gender and Society*, 1, 125–51.

Wilkins, S. (2001) 'Ageing, chronic illness and self-concept: A study of women with osteoporosis', *Journal of Women and Aging*, 13, 73–92.

Williams, S.J. & Bendelow, G. (1998) *The Lived Body: Sociological Themes, Embodied Issues* (New York: Routledge).

Winterich, J.A. (2007) 'Aging, femininity and the body: What appearance changes mean to women with age', *Gender Issues*, 24, 51–69.

3
Reconceptualising Later Life: Using Qualitative Methods to Refine Understanding of New Ageing Populations

Karen Lowton

Introduction

In developed nations, 'later life' is traditionally conceptualised as a time lived past a fixed retirement age, typically beginning in the seventh decade of the lifespan and lasting for around 20 years. As increased longevity continues, with cohorts experiencing better health through higher standards of living, a growing number of people are now living as nonagenarians and centenarians. The notion of the age at which 'later life' begins is therefore being pushed back further, such that it is now unusual for individuals in their sixties to self-identify or act as 'old' people.

In a similar fashion, the remarkable progress in medical treatment and care over the last four decades has provided many benefits for those who historically experienced limited life expectancy and profound disability. However, medical progress has rarely provided a 'cure'. Instead, through raising the average life expectancy of certain groups, more issues have become pertinent in the context of ongoing treatment and its impact on social experiences and identity in mid- and later life. There is currently very little understanding about how later life is conceptualised by these 'new' ageing populations; the joys and frustrations of living into 'old age', or how society can work to improve both life quantity and quality for these groups.

Parallels can be drawn here to the lives of people in other non-traditional groups and the uncertainties they face. Giddens (1992) draws on the example of lesbian women who needed to find the rules for a relationship that did not have an equivalent set of rules to a heterosexual one. At the leading edge of new lifestyles, he calls them 'prime everyday experimenters' (1992, p. 135), shaping their lives as they want to live them, and paving the way for others to follow. Giddens argues that, given the changes to contemporary society and the shifting role of relationships, the work done by lesbian women is also becoming more commonplace for hetero-sexual couples. Following this idea, the pioneering work of new ageing populations could be taken up by more traditional groups in future. This chapter seeks to highlight the uncertainty over how far members of new ageing populations have the potential to become 'everyday experi-menters' who exercise free choice in their lifestyles as they grow older, and to what extent these populations will age more through constraint; by living a life influenced by what health will allow and what main-stream society and its services offer in response. The chapter begins by examining the nature of the groups that make up the new ageing popu-lations and the drivers behind their improving longevity and quality of life. In understanding where qualitative methods can begin to address some of the new research agendas for these populations, four key areas of enquiry are considered. Firstly, how health and social care needs can be identified and met as individuals age; secondly, what the roles of family and friends might be in supporting individuals as they grow older; thirdly, how far identity and social roles are shaped both by the underlying condition and society's responses to it; and finally, how the 'older' person of the future may not be classified usefully by chronological age. I will elaborate these four areas by focusing on people with cystic fibrosis (CF, an inherited genetic condition) and the research colleagues and I have undertaken with this group. The chapter concludes by exploring how qualitative research can help understand these issues across the lifecourse of new ageing popu-lations, and how often sensitive areas can be addressed in research practice.

What are the new ageing populations?

Better understanding of disease processes has led to significant advances in preventative, routine and curative medical and surgical techniques. Marked changes in social attitudes and the social contexts in which health interventions are delivered have enabled many people to live a healthier later life than that which was possible just a few decades ago. Conversely, unintended consequences of biomedicine have led some people to live their life very differently to the one that was initially envis-aged for them. These developments have brought into existence the new ageing populations, whereby those born with serious health conditions who did not previously survive to adulthood and those developing pre-viously life-limiting conditions in early or mid-life now routinely reach mid- to late adult life.

New ageing populations essentially comprise three main groups, although these are not necessarily mutually exclusive. Firstly, through a fuller understanding of the genetic basis of conditions and refinement of medical and surgical interventions, survival age and quality of life have improved dramatically for many of those who traditionally died in childhood from conditions such as CF, congenital heart disease, and Down's syndrome (see Chapter 5). Through using disease registers we know that most infants born with these types of conditions today can now expect to live at least into mid-adulthood; for example life expectancy for those with Down's syndrome is now around 56 years (Glasson et al., 2002), with average survival for newborns with cystic fibrosis predicted to approach 50 years (Dodge et al., 2007).

Secondly, continuing advances in medicine and technology have enabled more people to survive the onset of life-threatening illness in childhood or mid-life, such as acute childhood leukaemia or HIV infection. Indeed, many young men who survived an HIV diagnosis in the mid-1980s are now celebrating their 60th birthday. An analysis of 22 cohorts of people with HIV in Europe, Canada and Australia indicated that this markedly increased longevity has primarily arisen through the introduction of highly active antiretroviral treatment (HAART), transforming the trajectory of HIV infection from an acute fatal illness to a manageable long-term condition (Cascade Collaboration, 2003).

Those affected by the 'failures' of regulation or unintended consequences of biomedical treatment make up the third group. For example, thalidomide was withdrawn in Britain in 1961 after significant numbers of infants were born with birth defects; there are now around 450 thalidomide impaired people in their late 40s experiencing increasing disability in the UK alone (Bent et al., 2007). More recently, technological developments in neonatal intensive care are enabling clinicians to rescue increasingly smaller and extremely premature newborns. Although their survival age is improving over time this is increasingly thought to be associated with a high prevalence of long-term physical and mental health problems, including sensory impairment, cerebral palsy, and learning difficulties (Saigal & Doyle, 2008).

The size of these three groups is continuing to grow in terms of types of underlying conditions involved, numbers becoming adult, and length and quality of their lives. Through better medical knowledge and understanding, new therapies are being developed that expand the range of possibilities for medical intervention. This not only increases the numbers of people able to survive different conditions, but also offers new opportunities for the survival of groups who would not previously have been seen as being beneficiaries of medical advances. New ageing populations are also growing through the transformation of the contemporary social world, where the existence of people with many different complex health conditions has raised concerns about how to ensure not only longer lives, but also how to sustain a better quality of life for all.

Although most of these individuals now live well into adult life, members of new ageing populations remain unlikely to survive in good health to a traditional pensionable age or beyond. Whilst increasing longevity is a very welcome development, living into a non-traditional later life, in terms of what chronological age can be reached, what health conditions may be experienced and what social opportunities are both available and achievable, raises many new issues as the position of individuals in family life and society are negotiated. Changes currently seen in the patterning and nature of contemporary 'old age' raise important questions for families, researchers, providers and policymakers alike about our understanding of the current circumstances and opportunities that surround ageing.

What are the research agendas for new ageing populations?

The epidemiological transition of new ageing populations mirrors to some degree the ageing of 'healthy' populations in terms of growth in total population size, the numbers of adults surpassing the numbers of children, the increased burden of disease and family support required, and the increased heterogeneity arising from ageing populations and social diversity. While some of the challenges that these factors present for the resourcing of adult health and social care have begun to be rehearsed, a more comprehensive understanding of the social context in which new populations are ageing is needed, together with an understanding of how this affects 'older' people's contemporary identities, roles and relationships.

Broadly, in providing care and support to members of new ageing populations, we need to ask 'for what sort of future late adulthood are young people being prepared?' Equally, in what ways does our current social context shape the identities, opportunities and experiences of new ageing populations as they grow older? What might the future hold for these groups, and the cohorts behind them? And what parallels can be drawn for all the above between new ageing populations and ageing populations at large? How can pioneers in one group provide ideas and paths to be used by others? Four key domains where qualitative enquiry can illuminate answers to these questions are in health and social care; relationships with family and friends; identity and social role; and uncertainty about the future.

Ageing, health and social care

People with complex conditions require care over a lifetime. Each group raises very different concerns of how the self and others define, perceive and manage health, illness and disability; the level of interaction, involvement and engagement with health professionals, assistive technologies and biomedicine; and the influence of varying dependencies, social networks and wider social contexts on the lives of individuals. They also raise questions of how biomedical and technological advances can be used to recognise, respond to and shape health needs and social roles.

With increased survival, and despite biomedical intervention, there is growing evidence of a significant and increasing prevalence of health deterioration, with challenges arising from chronic health impairment. Indeed much health maintenance is currently viewed through the lens of 'coping' with a condition (Abbott et al., 2001). Health deterioration may be due to the underlying disease process itself, the ways in which the condition is treated, an accelerated or early ageing process, or ageing following an early insult to health. Whatever the cause, the health and treatment burden in adulthood may be far greater for new ageing populations than that experienced by the general population. Even where medical treatment is expected to be minimal in later life, such as for those surviving childhood cancer, the range and scope of surveillance to monitor health is increasing. Here similarities can be drawn to adults living with chronic, often multiple, conditions.

For a minority of these new ageing groups, disease databases and death registrations allow quantification both of the improvements achieved and inequalities encountered in survival age and health status over time. For example, during 1999–2007 a national (then European) database for those treated for CF allowed both economic and clinical evaluations of standard and new therapies or approaches to care. Recently, an annual cross-sectional analysis of all deaths from CF in England and Wales from 1959–2008 has found that those in a higher socioeconomic group are more likely to die above the median age of death than those from a lower social class (Barr et al., 2011). In a similar manner, disease registers provide a basis for quantitative research and allow the incidence, prevalence and course of conditions to be tracked, thereby offering scientists opportunities to improve further treatment and care. However, although registers and databases are useful in planning services and measuring progress (or its absence) on a national scale, they cannot be used to understand how people access, use, leave and experience formal health and social care services in mid- to later life, or in what ways formal services are able to respond to the many expressed and hidden needs of patients and their families.

Increased survival and older age has brought clinical recognition of a higher prevalence and potential burden of condition-related morbidities, such as CF-related diabetes, osteoporosis and incontinence (Parkins et al., 2011); and increased risk of dementia both in those ageing with Down's syndrome (Wisniewski et al., 1985) and HIV (McArthur, 2004). Here, the growth in the numbers of ageing individuals is important, as many disorders appear only to be clinically recognised and treated when a new age milestone has been reached for a significant number of that group, or when circumstances surrounding the disorder start presenting new dilemmas. Furthermore, these age-related conditions often occur earlier in the lifespan than for the general population, and can cause significant disruption to daily routines. For example, the occurrence of urinary incontinence in young women with CF only began to be noted in the clinical literature at the beginning of this century. Although the physical problems facing incontinent young women with CF in their early twenties are likely to mirror the ageing-related problems of much older women in wider society, we know little about the psychosocial impact for these young women, how they and their clinicians raise and interpret the growing need for new interventions as this group age and experience new disorders, or how strategies can be formulated that may improve further their quality of life. There is also little evidence to suggest that knowledge is shared between clinicians who treat women with urinary incontinence post child birth, as this group is closer in age to young women with CF.

For those reaching mid-life with chronic HIV infection, the emphasis in medical management has similarly shifted from treating often fatal opportunistic infections to managing the consequences of lifelong treatment and associated conditions. Just 30 years ago those diagnosed HIV positive were usually given less than three years to live. If they survived, another fatal prognosis was soon delivered. Although we have a wealth of clinical quantitative data to map improvement in life quantity and quality, we know very little about how individuals live their lives through regular warnings that death is imminent, or what strategies and support are available to those who gave up work and spent their savings, only to find that life continues. The first cohort of 'survivors' are now approaching a traditional retirement age, but it is not known whether their health and social care needs in later life can be met in a way that is both holistic and personalised. For example, Mike Youle, director of the HIV Training and Resource Initiative, once asked, 'should we be building nursing homes for old queens?'

Both advances in medical interventions and the subsequent appearance of new conditions draw attention to the many social and ethical uncertainties surrounding further life extension in both current and emerging new ageing populations. 'Right to life' debates have focused traditionally on the early years of life for those with complex conditions, and have not really addressed the uncertainty surrounding the longer-term survival of these now adult populations. In the context of organ repair and transplant for example, clinicians were at first hesitant to perform surgery on individuals where the outcome was uncertain. It was rare before the 1970s for a British infant born with Down's syndrome, congenital heart defect and cognitive impairment to be selected for life-saving cardiac surgery (Silverman, 1985). Today, these infants routinely undergo early corrective surgery and medical care. However, despite most transplant surgery having moved from the experimental to the routine, there is no guidance on whether offering a heart transplant to an adult with Down's syndrome (Leonard et al., 2000) would be the 'right' decision to make. Similarly, the first lung transplant for a young person with CF was not performed until 20 years after the first lung transplant had taken place, for fear that CF would infiltrate the donated lungs. Today, adults with CF are requiring and surviving lung transplantation in ever larger numbers, yet it is not known how the uncertainties of later life post transplant are managed by both clinicians and their patients, or whether there is a limit to the number or type of interventions that can be performed. In a different vein, recent debate has extended to the place of a 'right to death' for those with severe disabling life-limiting conditions such as motor neurone disease (Curtice & Field, 2010). Although the 'right to death' sets up new dilemmas for those providing any type of care or support to these populations at the end of their life, we have not yet got to grips with the personal, professional and social impact of new ageing populations at each stage across the lifespan before its end (see Chapter 7).

As biomedicine, the lifecourse itself and its associated health needs become more complex, diverse new challenges and opportunities for different professional health care groups are also created, and destabilise assumptions of the traditional 'health professional'. How far health professionals will be forced to abandon their care assumptions based on historical age norms, and relate instead to a much more diverse set of patients who will require treatment and care for both condition- and adult age-related disorders is not yet clear. We need a much fuller understanding of the purpose and meaning of care for these groups; this includes not only for what lifecourse trajectory adults are being prepared, but also to what extent health professionals can assist individuals in achieving their social goals.

The optimal setting required to provide complex medical care to these new ageing populations is additionally poorly understood, with many health services continuing to be delivered through 'traditional' age- and hospital-based services. Understanding of how the multitude of specialist adult services can be configured and negotiated in order for 'seamless care' to be experienced past paediatric care is in its infancy. Transition in this context is usually understood in terms of moving individuals from specialist paediatric to adult services where, despite care being delivered by a multidisciplinary team from a variety of professional backgrounds, the focus generally remains on maintaining optimum physical health, rather than considering wider psychosocial needs. Although transition services are now available for many young people leaving paediatric care, how care delivery in mid–late adult life is anticipated and experienced by these new ageing populations is again largely unknown. Through qualitative interviews with parents and partners of those attending adult health services, I found that family members feel they are usually considered by health professionals as a support during periods of acute illness or crisis (Lowton, 2002), rather than being an integral and ongoing part of an adult's social wellbeing in later life. In our work we argued for the centrality of the notion of family care and support given throughout the lifecourse of these new ageing groups and uncover new tensions in family relationships with adult health care professionals (Iles & Lowton, 2010). It appears that families may become disillusioned with health care professionals at the time of transition, perhaps through the confidentiality of the adult patient's medical care leading to close parent-provider bonds being broken, or through professionals believing that family expertise is no longer required in adult care. It is important to investigate whether there are optimum models for family or other non-professional care and support roles within formal adult services, how these may be modified for different groups, and what factors might explain the diversity of potential options for care.

Ageing, the family and friendships

Despite acknowledging the expertise that parents possess, one recent clinical opinion of parents of adults with congenital heart disease views them as 'carers emeritus' when their child reaches adult services age:

> As clinicians, remember that parents are co-experts in this process and are transitioning into a 'retirement' or 'caregiver emeritus' role (Ross & Fleck, 2007, p. 813).

On the contrary, parents are likely to have many different roles in continuing to support their ageing offspring: the emphasis may remain on caring as defined in more traditional terms, but roles are also likely to encompass providing practical and emotional support through guidance and advice, and to feature a more organisational component. We know that for those affected by CF the nature and position of parental care and support in the ageing of new adult populations is complex (Iles & Lowton, 2010), and relationships between parents and health professionals can be strained (Lowton, 2002), but what exactly leads some families to continue to enjoy productive relationships with adult health and social care services, whereas others feel suddenly excluded or 'made redundant' at transition is unknown. What are the developing norms of family support and other informal caring mechanisms for newly ageing adults who are seen within adult services? How are changes or interruptions to social roles negotiated between different family members? How, if at all, can these roles be conveyed to health and social care service professionals?

Although health professionals continue to coordinate the bulk of formal medical care, previously complex interventions which over time have become 'routine' treatments have been moved out of the hospital and into the home environment. Families are gradually taking over many of the traditional caring roles performed by hospital nurses, with the home becoming more like another clinical environment. This shift has enabled 'patients' to participate more fully in family and social life by limiting the disruption that hospital treatments can impose. However, home-based treatment has the potential to restrict the social world and result in greater isolation for family carers, through added responsibility for care delivery and surveillance. As Gillies et al. (2003) argue in the context of 'healthy' children, transitions to adulthood and freedom are experienced by all members of the family, who may lend very different interpretations to the process. In supporting members of the new ageing populations, it is vital to know more about how family members are able to gradually hand over home treatment responsibility to their offspring as they grow older, how far young people can maintain responsibility for their own treatment as they age, and in what ways different care environments affect their capacity to do this.

Understanding the experience of growing older with novel, chronic conditions cannot be complete without acknowledging the family setting as an ongoing process which constitutes and reconstitutes meaning (Gregory, 2005). This is even more pronounced at a time when more traditional family structures and living opportunities are also undergoing significant change. For example, young children with Down's syndrome historically entered institutional care. Today, these young people are more likely to live at home with their family and aspire to supported community living when older, a life-changing transition full of complexity and uncertainty (see Chapter 5). On the other hand, those growing older with HIV and experiencing deteriorating health may desire some form of non-traditional supported living, as noted by a 70-year-old man living with haemophilia and HIV (BBC Radio 4, 2006):

> *And I think probably what worries me more than haemophilia or HIV or any of those things is old age itself, and what happens to you. I don't want to go into one of these homes – it's unlikely that any home would ever accept me anyway, not with all my viral problems and haemophilia as well.*

In the absence of a script on how to age prematurely and how to deal with (perceived) social barriers to traditional ageing facilities, new norms of living and for planning have to be found. Like less traditional groups planning care in later life, such as lesbian, gay and bisexual older people (Almack et al., 2010), some members of the new ageing populations may be able to create new caring mechanisms and relationships which could potentially pave the way for other groups who need to plan informal care in a non-traditional family context. In providing social support to new ageing populations health and social care services need to understand the internal tensions and collaborations that families experience in their attempts to enable independence (or interdependence), and acknowledge the reliance on family and other informal support and care throughout the lifespan. For example, how are goals negotiated and tensions managed between parents preparing young adult offspring for independent living and supporting them in the context of their complex conditions? How do adults and their families negotiate and manage steps to independent living, and what social support systems are in place when independence is no longer achievable or desired?

In thinking of wider social networks, one key to beating depression in later life for those who are ageing 'healthily' appears to be to maintain friendships (Cornwell & Waite, 2009). Many people with a rare or complex condition form friendships with others in the same position. But how far is that sustainable for those who have CF or chronic HIV infection? As they age, these individuals witness the deaths of most of their contemporaries, so that in mid-life new friendships with those with the same condition may be unwanted. Very little is known about close friendships within these populations and how these can be supported, especially at end-of-life, which itself is likely to be made more complex against a medical backdrop of actively seeking life extension for the cohort.

Ageing, identity and social role

A further important dimension regarding the emergence of new ageing populations is how individuals perceive and construct their own identity as they age. For many of these groups a primary identity has been created early in the lifecourse by the underlying biomedical condition and its interface with health and social services. However, qualitative investigation suggests that many young adults with CF, despite having a life-shortening genetic disease, view themselves to be in a liminal state, neither 'healthy', 'ill' nor 'disabled' (Lowton & Gabe, 2003). Further investigation of how individuals construct their self-identity and interpret their health status may help those supporting them, both professionally and informally, to understand the challenges and aspirations to which new ageing populations aspire.

Questions also arise of identity in mid- and late life, both in terms of self-identity and other defined identity, and how these identities, either maintained through the lifecourse or threatened through ageing, might influence the desire or ability to reach and maintain independence. Furthermore, few of these groups have a 'tradition' of disclosing their condition, particularly when some conditions, such as CF, can be much less apparent to an 'uneducated' public than others. Disclosing can be fraught with uncertainties, for example in timing *when* to tell and weighing up the benefits and drawbacks of *whom* to tell (Lowton, 2004). This can be challenging enough for individuals who are aware of the various implications of decisions because an establishment of adults have already gone through the same experiences in mid- or later life. However, for many groups these adults are the *first* of their kind in these situations and possess no experiential knowledge of the best path to take.

Reaching mid-life also raises new questions surrounding conceptualisation and management of biography and social positioning in uncharted terrain, including educational achievement and employment opportunity, romantic relationships, and safe housing. Also of interest to qualitative researchers are the limits of social roles, and how worries about future health might affect current and anticipated social positions. However these issues are only now beginning to be addressed, for example, in how teenagers' fears over future infertility and possible early death from CF may shape their forming of romantic relationships (Iles & Lowton, 2008).

Equally, we need to find out whether and how far society can continue to modify traditional educational and working practices to accommodate those who currently experience periods of acute debilitating illness or incapacity. Here, qualitative work is needed in order to understand how opportunities across the lifecourse can best be structured to enable early gains in social roles to be sustained as individuals age. For example, it is important to consider the concept of 'serious leisure', the focused participation in a substantial activity through which people gain satisfaction through learning and using new skills and knowledge (Stebbins, 1982). Patterson and Pegg (2009) used qualitative interviews to understand how those ageing with learning disabilities are able to use their leisure activity as a form of 'work', with clear aims and objectives, commitment and time utilisation. Building on this, we also need to understand more about how virtual environments can shape and maintain identity for members of new ageing populations. For example, by constructing a new identity through an avatar, Jason, a wheelchair-bound 32-year-old with multiple disabilities was reported to play for around 80 hours per week in the online game *Star Wars Galaxies*. Jason noted that online, for the first time in his life, his 'virtual' peers treat him as an equal:

> The difference between me and my online character is pretty obvious. l have a lot of physical disabilities but in Star Wars Galaxies I can ride an imperial speeder bike, fight monsters, or just hang out with friends at a bar. ...The computer screen is my window to the world. Online it doesn't matter what you look like. Virtual worlds bring people together – everyone is on common ground...The internet eliminates how you look in real life, so you get to know a person by their mind and personality (Cooper, 2007, p. 14).

Shaping the future

The relative newness of many biomedical interventions and the short-term, health-focused outcomes that are currently measured, predominantly by quantitative methods, mean that the longer-term outcomes and limits of medical progress are yet to be understood. Many members of the new ageing populations now occupy a wide range of roles, including student, employee, parent, and partner. However, there is likely to be much variation on how well entry to and maintenance of these roles can fit with the competing health and care needs of ageing individuals. Only a tentative ontological security exists therefore, influenced by a centrality of uncertainty for members of new ageing populations, their families and professionals alike. Through questionnaire surveys we know that many individuals report needing further information about their future health and its unpredictability (see for example Sawicki et al., 2007), and about the psychological effects of growing older with often very complex and relatively novel physical impairments (see for example Kennelly et al., 2002). More widely, many adults with child-hood conditions express uncertainty in making any long-term plans. For example, in the context of adults with congenital heart disease, Claessens et al. (2005) found areas of difficulty in financial decision-making includ-ing in relation to employment choices, spending and saving, and pension provision. This lack of ontological security is likely to be further under-mined by the unpredictability of the disease process as individuals age, and by new technological interventions that may hold only a fragile promise of success.

There is also a pressing need to widen debates on the interconnections between health, identity, disability and ageing in the emergence of new ageing populations. Their growing lifespan throws light on important dimensions of the current interface between biomedicine and society, as well as the way in which notions of ageing and traditional old age are being challenged and reconfigured (see Chapter 1). Attempting to delin-eate how a number of groups with particular conditions might constitute a category of new ageing populations draws the immediate response as to whether they share any intrinsic connections at all, or whether there are any particular boundaries which might separate them out from other groups. The answer to both these questions is 'yes'; furthermore the 'new' in new ageing populations is an aspect of the changing nature of ageing itself as well as representing novel developments in biomedicine. Discussion in gerontology of 'normal' ageing has had to adapt over the past few decades to the fact that such an overarching idea of 'normal' no longer exists (Jones & Higgs, 2010). Longevity is increasing in the major-ity of the world's nations, and the epidemiology of old age is being trans-formed as many conditions become modifiable, with later ages of onset. As 'normal' ageing becomes less definable so too does the concept of old age which can no longer be encompassed as a residual 'social category' of those whose dependency and infirmity mark them out from mainstream society. To varying extents new ageing populations share in this new configuration of ageing rather than being separated from it. They may not wish to be considered old, but they are positioned within the same processes and dynamics as those who are. Because new ageing popu-lations have benefited from the developments and unintended problems that have emerged from biomedicine they may share the same relation-ships with biomedicine as do the chronologically older group. Addition-ally, new ageing populations want to see their identities and interests acknowledged and their independence maintained in their interactions with health and social care professionals. In this way new ageing popu-lations are an emergent aspect of the potentiality of biomedicine, and of the changing nature of social relations surrounding these developments.

Understanding more about contemporary later life through qualitative research methods

Unpacking the process

By using qualitative research methods much insight can be gained into contemporary ageing-related areas in our society where very little is currently known. In-depth interviews in particular lend themselves well to understanding more fully the processes that lead to the achievements and frustrations of new ageing populations. Analysis of qualitative data gathered allows the opportunity of understanding the diversity of the 'later life' of populations who are currently unlikely to age beyond their sixth decade, and enables researchers to reconfigure accurately what later life means in contemporary societies. Interviewing people is a privilege; researchers are let into people's lives, and interviewees open up to a stranger. If social researchers want to advance knowledge of ageing populations they need to be prepared to be open to new ideas as to how other people experience their lives and what is important to them.

Interviewing individuals at different stages of the lifecourse, for example teenagers, young adults, those who have survived past the average survival age for the group, and those who have undergone the last medical procedure that is available to treat them, brings focused insights into how key lifecourse events are shaped. In interviewing people of different ages with CF, my colleagues and I were able to gain an understanding of how transition to adult health services affects teenagers' perceptions of their relationship with their parents (Iles & Lowton, 2010) and brings worries of their future to the fore (Iles & Lowton, 2008); how ageing, the disease process and biomedicine shapes adults' perceptions of their health (Lowton & Gabe, 2003); how decisions regarding organ transplantations are shaped (Lowton, 2003); and how familiarity with place and professionals serve to influence parental experiences of their offspring's end of life care (Lowton, 2009). However, this just taps the tip of the iceberg; additional challenges for new ageing populations to negotiate, and for society to more fully understand, include how identity is constructed and evolves, and how achievement of a wide and diverse array of social roles can be advanced and maintained. Furthermore, as the reach and scope of medical interventions continue to widen, and quality and quantity of life improves further, questions of future intervention expectations and social roles additionally arise.

Of course, at every life stage many different people invest in and are affected by the lives of members of new ageing populations, from parents and partners to peers and professionals. Consideration of family and other supporters' perceptions and needs with regard to members of new ageing populations are only now beginning to be investigated. Yet considering more fully particular issues from different perspectives allows us to tri-angulate the data in order to obtain a more comprehensive account of the nuances of the circumstance. For example, interviewing the parents and partners of young adults with CF highlights how transition to adult care and end of life care affects not only young people but also the family's relation-ships with their child and health care professionals (Lowton, 2002, 2009).

Uncovering the processes and effects of ageing and illness in these populations can be achieved by employing many different qualitative approaches to data collection, particularly when the issues for enquiry may be deemed to be sensitive. For example, letters and self-recorded interviews have allowed researchers to gain valuable insights from those who may be unable or unwilling to meet a researcher or take part in a focus group interview. Grinyer (2002), for example, explores through parents' written and tape recorded accounts, the sexuality, fertility, per-sonal finances and management of conflict of young people with cancer, along with the disease's effect on family dynamics. This method, along with face-to-face in-depth interviews, has also been used successfully to understand more fully how end-of-life care for young people with CF, who are both striving to maintain normality and be 'well', yet are dying, is experienced by their parents (Lowton, 2009). Additionally, participant-generated video diaries have been used successfully with young men with Duchenne muscular dystrophy to capture more fully the social world and self-identities of this new ageing population (Gibson et al., 2007) yet are not without their challenges (Gibson, 2005).

Researching sensitive areas

As the section on ageing, identity and social role argued, medical con-ditions affect peoples' self-identity. Extremely sensitive issues such as nego-tiating disclosure of a complex condition, or imagining worsening health or dying at a relatively early age, are likely to be key elements of psycho-social life, and deserve attention to capture them more fully. Sending potential participants the research topic guide together with the invita-tion to participate and information about the project enables respondents to be more informed about the nature of the research. It also allows them time to think about what they want to recount, especially if subjects under discussion will involve events that took place long ago.

As most qualitative research conducted with those ageing with a complex health condition and their supporters can be deemed to be sensitive consideration of ethical issues quite rightly comes to the fore. However, our hesitance in involving people in research who may be deemed 'vulnerable' may mean that we lose valuable opportunities to understand more fully how the lifecourse is experienced by these newly ageing families. Adapting traditional qualitative interview methods and asking people to tell their story can allow for an insightful research interview and a cathartic experience for those taking part. The latter is often ignored in the discussion about the impact of qualitative research on participants, as the focus is generally on the potential negative impacts rather than the positive ones (e.g. for ethical approval). For example, in a study of bereaved parents of young people with CF (Lowton, 2009), the majority of the mothers remarked that they had spent years wanting to talk more to their friends about their child, but these friends had often acted in ways that closed down the conversation or otherwise made it difficult for them to talk. Deciding to participate in that study was for some a very difficult decision, with potential respondents weighing the opportunity to talk more about their child against remembering the exceptionally difficult time in their life that they had attempted in part to forget. The excerpt of the letter below was sent to me about a month after a mother of two children, both of whom had died from CF, had participated. One of her children had had what she reported to be a 'good death'; the other had not, prompting her to become upset at many points during her interview. She went on to write:

> *'It is very strange, Karen, but since you have been down, and I know I must have talked a load of rubbish, as I have so much in my heart, and mind, and when it comes to it, it always comes out (to me, anyway), a load of rubbish, but once again since you have been down I feel so much better, whether it was that you were a stranger, I don't know, but I feel, much lighter, if that makes sense?'*

However cathartic an interview may be for a participant, for many of these qualitative projects it is essential to have a strategy in place to deal with distressing information, both for the respondents and members of the research team (see Chapter 2). Being very careful to end sensitive or distressing interviews on a conversational and positive note (for example asking a bereaved parent 'can you tell me a funny thing that you remember your child doing?') serves to ensure that respondents are able to remember the more positive or treasured aspects of their child's life at the interview's conclusion, and can give the researcher a more rounded picture of what a 'normal' life the young adult had managed to achieve, if only in part.

Of note, the person transcribing these interviews may listen to the same conversation without ever meeting the person who may be recounting distressing information or witnessing the context in which the interview was conducted. Being mindful of these often isolated workers, and enabling them space to talk about their reactions to interview material is a crucial part of the research process, but one that is not usually acknowledged.

Conclusion

Increasing medical progress has enabled both improved life expectancy and movement of many once marginalised populations into the mainstream of society in terms of socialisation, education and employment. By using qualitative research methods we can more fully understand and help further this progress, previously defined through a narrow biomedical model.

On the face of it, new ageing populations comprise a large number of disparate groups. For example, conditions may be present at birth or acquired in mid-life, and medical intervention may consist only of surveillance, through to reliance on regular treatment and admissions to hospital. However, these groups share between them a multitude of common challenges, such as uncertainty over when to disclose an underlying condition, social roles that can be threatened by failing health, and ongoing reliance on informal care or support. For example, the need to be supported in later life by informal carers, who may not necessarily be family members, appears to be shared by 'ageing everyday experimenters' (Almack et al., 2010), some more traditional 'older' people, and members of new ageing populations dying in their thirties and forties. Reaching mid-life for new ageing populations raises new questions surrounding conceptualisation and management of identity, biography and social positioning in uncharted terrain, and provision of health and social care in later life.

There is now a greater realisation that what actually constitutes 'ageing' and 'health' is becoming more difficult to define and demarcate, whether in terms of physiological 'normality', cultural expectations of 'older' people, or social provision for those growing older. This is partly because of the commonalities that many groups share, regardless of their chronological age. Qualitative research, often criticised for not being able to generalise specific findings from its focus on small groups of people, is an ideal approach to use in uncovering both common concepts across these groups, and more unique aspects of ageing, together with possible solutions to the challenges they face. Through working in this way we can further increase our understanding of how contemporary ageing is experienced in Western societies, vital if we are to shape future policy and practice to improve the circumstances of these new ageing populations.

Annotated further reading

Beynon, C.M., Roe, B., Duffy, P. & Pickering, L. (2009) 'Self-reported health status, and health service contact, of illicit drug users aged 50 and over: A qualitative interview study in Merseyside', *United Kingdom BMC Geriatrics*, 9:45 doi:10.1186/1471–2318–9–45.

Through interviews with ten older illicit drug users, an understanding of health status and service contact is beginning to be uncovered; an important aspect of understanding contemporary social life as populations age.

Gibson, B.E., Young, N.L., Upshur, R.E.G. & McKeever, P. (2007) 'Men on the margin: A Bourdieusian examination of living into adulthood with muscular dystrophy', *Social Science & Medicine*, 65 (2007) 505–17.

Through semi-structured interviews and video diaries Gibson et al draw from Bourdieu's critical social theory to explore the identities and social position of ten young men with Duchenne muscular dystrophy.

Patterson, I. & Pegg, S. (2009) 'Serious leisure and people with intellectual disabilities: Benefits and opportunities', *Leisure Studies*, 28 (4) 387–402.

Through qualitative interviews with ten clients with mild and moderate intellectual disability, the theoretical notion of serious leisure is more fully understood in the context of those with intellectual disability.

References

Abbott, J., Dodd, M., Gee, L. & Webb, K. (2001) 'Ways of coping with cystic fibrosis: Implications for treatment adherence', *Disability and Rehabilitation*, 23 (8) 315–24.

Almack, K., Seymour, J. & Bellamy, G. (2010) 'Exploring the impact of sexual orientation on experiences and concerns about end of life care and on bereavement for lesbian, gay and bisexual older people', *Sociology*, 44, 908–24.

Barr, H.L., Britton, J., Smyth, A.R. & Fogarty, A.W. (2011) 'Association between socioeconomic status, sex, and age at death from cystic fibrosis in England and Wales (1959 to 2008): Cross sectional study', *British Medical Journal*, 343:d4818 doi: 10.1136/bmj.d4818.

BBC Radio 4. The Archive Hour. (2006) 'Living with Aids: Blood Brothers', http://www.bbc.co.uk/radio4/archivehour/pip/t2ihj/ (accessed 20.03.2011).

Bent, N., Tennant, A., Neumann, V. & Chamberlain, M.A. (2007) 'Living with thalidomide: Health status and quality of life at 40 years', *Prosthetics and Orthotics International*, 31, 147–56.

Cascade Collaboration (2003) 'Determinants of survival following HIV-1 sero-conversion after the introduction of HAART', *Lancet*, 362, 1267–74.

Claessens, P., Moons, P., de Casterle, D., Cannaerts, N., Budts, W. & Gewillig, M. (2005) 'What does it mean to live with a congenital heart disease? A qualitative study on the lived experience of adult patients', *European Journal of Cardiovascular Nursing*, 4, 3–10.

Cooper, R. (2007) *Alter Ego: Avatars and their Creators* (London: Chris Boot Ltd).

Cornwell, E.Y. & Waite, L.J. (2009) 'Social disconnectedness, perceived isolation, and health among older adults', *Journal of Health and Social Behaviour*, 50, 31.

Curtice, M. & Field, C. (2010) 'Assisted suicide and human rights in the UK', *The Psychiatrist*, 34, 187–90.

Dodge, J.A., Lewis, P.A., Stanton, M. & Wilsher, J. (2007) 'Cystic fibrosis mortality and survival in the UK: 1947–2003', *European Respiratory Journal*, 29 (3) 522–6.

Gibson, B.E. (2005) 'Co-producing video diaries: The presence of the "absent" researcher', *International Journal of Qualitative Methods*, 4, 4. http://ejournals.library.ualberta.ca/index.php/IJQM/index.

Gibson, B.E., Young, N.L., Upshur, R.E.G. & McKeever, P. (2007) 'Men on the margin: A Bourdieusian examination of living into adulthood with muscular dystrophy', *Social Science & Medicine*, 65 (2007) 505–17.

Giddens, A. (1992) *The Transformation of Intimacy* (Cambridge: Polity).

Gillies, V., Holland, J. & Ribbens McCarthy, J. (2003) 'Past/present/future: Time and the meaning of change in the "family"' in G. Allan & G. Jones (eds) *Social Relations and the Life Course* (Basingstoke: Palgrave Macmillan).

Glasson, E.J., Sullivan, S.G., Hussain, R., Petterson, B.A., Montgomery, P.D. & Bittles, A.H. (2002) 'The changing survival profile of people with Down's syndrome: Implications for genetic counselling', *Clinical Genetics*, 62, 390–3.

Gregory, S. (2005) 'Living with chronic illness in the family setting', *Sociology of Health & Illness*, 27 (3) 372–92.

Grinyer, A. (2002) *Cancer in Young Adults: Through Parents' Eyes* (Buckingham: Open University Press).

Iles, N.J. & Lowton, K. (2008) 'Young people with cystic fibrosis' concerns for their future: When and how should concerns be addressed, and by whom?', *Journal of Interprofessional Care*, 22 (4) 436–8.

Iles, N.J. & Lowton, K. (2010) 'What is the perceived place and nature of parental care for young people with cystic fibrosis as they enter adult health services?', *Health and Social Care in the Community*, 18 (1) 21–9.

Jones, I.R. & Higgs, P.F. (2010) 'The natural, the normal and the normative: Contested terrains in ageing and old age', *Social Science & Medicine*, 71 (8) 1513–19.

Kennelly, C., Kelson, M. & Riesel, J. (2002) *Thalidomide-Impaired People: Quality of Life* (London: College of Health and The Thalidomide Society).

Leonard, H., Eastham, K. & Dark, J. (2000) 'Heart and heart-lung transplantation in Down's syndrome: The lack of supportive evidence means each case must be carefully assessed', *British Medical Journal*, 320, 816–17.

Lowton, K. (2002) 'Parents and partners: The role of lay carers in the treatment and care of adults with cystic fibrosis', *Journal of Advanced Nursing*, 39 (2) 1–8.

Lowton, K. (2003) 'Double or quits: Perceptions of organ transplantation by adults with cystic fibrosis', *Social Science and Medicine*, 56 (6) 1355–67.

Lowton, K. (2004) 'Only when I cough? Adults' disclosure of cystic fibrosis', *Qualitative Health Research*, 14 (2) 167–86.

Lowton, K. (2009) '"A bed in the middle of nowhere": Parents' meanings of place of death for young adults with cystic fibrosis', *Social Science & Medicine*, 69 (7) 1056–62.

Lowton, K. & Gabe, J. (2003) 'Life on a slippery slope: Perceptions of health in adults with cystic fibrosis', *Sociology of Health & Illness*, 25 (4) 289–319.

McArthur (2004) 'HIV dementia: An evolving disease', *Journal of Immunology*, 157 (1) 3–10.

Patterson, I. & Pegg, S. (2009) 'Serious leisure and people with intellectual disabilities: Benefits and opportunities', *Leisure Studies*, 28 (4) 387–402.

Parkins, M.D., Parkins, V.M., Rendall, J.C. & Elborn, S. (2011) 'Changing epidemiology and clinical issues arising in an ageing cystic fibrosis population', *Therapeutic Advances in Respiratory Disease*, 5(2) 105–19.

Ross, H.M. & Fleck, D. (2007) 'Clinical considerations for allied professionals: Issues in transition to adult congenital heart disease programs', *Heart Rhythm*, 4 (6) 811–13.

Saigal, S. & Doyle, L.W. (2008) 'An overview of mortality and sequelae of preterm birth from infancy to adulthood', *Lancet*, 371, 261–9.

Sawicki, G.S., Sellers, D.E., McGuffie, K. & Robinson, W. (2007) 'Adults with cystic fibrosis report important and unmet needs for disease information', *Journal of Cystic Fibrosis*, 6, 411–16.

Silverman, D. (1985) 'Agreeing not to intervene: Doctors and parents of Down's syndrome children at a pediatric cardiology clinic' in D. Lanc & B. Stratford (eds) *Current Approaches to Down's Syndrome* (London: Holt, Rinehan & Winston).

Stebbins, R. (1982) 'Serious leisure: A conceptual statement', *The Pacific Sociological Review*, 25, 251–72.

Wisniewski, K.E., Wisniewski, H.M. & Wen, G.Y. (1985) 'Occurrence of neuropathological changes and dementia of Alzheimer's disease in Down's syndrome', *Annals of Neurology*, 17 (3) 278–82.

Part II

Under-Researched Ageing Populations

4

Doing Pakistani Ethnicity the Female Way: Issues of Identity, Trust and Recruitment when Researching Older Pakistani Muslims in the UK

Maria Zubair, Wendy Martin and Christina Victor

In this chapter we reflect upon our experiences of undertaking qualitative fieldwork with older Pakistani Muslim women and men living in the United Kingdom (UK). The significant increase that is expected within the next 20 years in the proportion of older people living in minority black and ethnic communities within the UK (Merrell et al., 2006) – particularly those living within the Bangladeshi and Pakistani communities (see Katbamna et al., 2002; Phillipson et al., 2003; Burholt & Wenger, 2003) – suggests a growing need for conducting research on this hitherto under-researched group of Bangladeshi and Pakistani older people (see Vincent et al., 2006; Victor et al., 2012). This is especially because of the particularly high levels of inequality and disadvantage experienced by members of these communities in the UK (Qureshi, 1998; Harding & Balarajan, 2001; Nazroo et al., 2004; Nazroo, 2006), and their higher levels of morbidity rates (Katbamna et al., 2002), which are likely to have important implications for how old age is experienced by members of these minority groups. However, as it is common with many other under-researched minority and migrant groups, doing qualitative research with older Bangladeshis and Pakistanis presents researchers with key challenges. These challenges stem not merely from the cultural and linguistic differences that may often exist between researchers (and also between the wider academic world) and these ethnic minority groups (see Boneham, 2002; Feldman et al., 2008; Hanna et al., 2008; Lloyd et al., 2008), but also relate to the particular social and cultural identities of the researchers *vis-à-vis* those they research. More specifically, how a researcher's various identities interact with those of their participants, and how a researcher is perceived by their participants can have an important influence on issues associated with access and recruitment (see Lee, 2008; Lloyd et al., 2008; Yip, 2008; Wray & Bartholomew, 2010). This is particularly the case where the ethnic and racial minority identity, or the socially marginalised status of participants may result in a mistrust, on their part, of perceived 'White officialdom' of the researchers, and hence the purposes and uses of the research (see Boneham, 2002; Levkoff & Sanchez, 2003; Moreno-John et al., 2004; Barata et al., 2006).

Drawing and reflecting upon our fieldwork for our ESRC *New Dynamics of Ageing* project: *Families and Caring in South Asian Communities*, we discuss in this chapter some of the issues of researcher identity, participant trust and the related challenges of access and recruitment that Maria (a Pakistani researcher in the project, and the first author of this chapter) faced in her capacity as a young Pakistani Muslim woman researching older Pakistani women and men. Our project aimed to explore the social identities, social networks and family lives of older people from South Asian communities and their own meanings and experiences of 'care' and 'support'. In accordance with its aim of exploring the subjective meanings and understandings of the participants, we made use of qualitative research methodologies – including semi-structured, in-depth, interviews and social network mapping. Our participants, who were a diverse group of mostly first-generation Bangladeshi and Pakistani migrants in the UK, and aged 48 years and over, were in most cases unfamiliar with the purposes and uses of academic research. Moreover, they varied in terms of the levels of their linguistic and literacy skills in English.

Given the variations in our participants' particular linguistic and literacy skills in English, we produced our information leaflets, consent forms and interview guides in Bengali and Urdu as well as the English language (see Zubair et al., 2010). We expected that conducting fieldwork in our Bangladeshi and Pakistani participants' own languages, facilitated by our own Bangladeshi and Pakistani female researchers, whose ethnicity 'matched' with those of our participants would allow for easier access into the social worlds of our participants, and also help with recruitment (see Bhopal, 2001; Boneham, 2002; McLean and Campbell, 2003; Gallagher-Thompson et al., 2006). However, as we shall illustrate in this chapter through Maria's experience of doing research with the Pakistani community, the issues relating to researcher identity and participant trust and recruitment are far more complex than a simple 'matching' of researcher-participant characteristics (see Levkoff & Sanchez, 2003; Yancey et al., 2006; Feldman et al., 2008; Wray & Bartholomew, 2010). The multiple and contextual nature of social and personal identities, and the heterogeneity of experiences within any ethnic minority group arising from internal differences in social class, gender, age, education, occupation, sexual orientation and so on makes such 'ethnic-matching' less useful than is often assumed (Twine, 2000; Wray & Bartholomew, 2010). The contextual and multiple nature of social and personal identities also reinforce the need for greater reflexivity in terms of how these multiple identities of the researchers and the participants interact and are played out within 'the field', and how this influences the research process (see Edwards, 1990; Bhavnani, 1991; Skeggs, 1997; Ramji, 2008). This is what we attempt to do in this chapter. More specifically, by focusing on the issues of access and recruitment within our own research, and using Maria's example, we aim to illustrate the interconnecting nature of a researcher's various identities – such as ethnic, gender and age identities. These interconnecting identities, we argue, make the researcher's status within 'the field' far more complex and dynamic with respect to the insider/outsider binary. We also point out that, while within our own fieldwork issues of participant access and recruitment remained closely linked with the issue of trust and rapport, the trust and rapport that developed between Maria and our participants was more related to Maria's embodied expression of her gendered Pakistani ethnicity, as opposed to a simple 'matching' of her ethnicity (and even gender) with those of our participants. In order to contextualise our fieldwork experiences, we first give below a brief description of our research project team and Maria's background.

Our research project team comprised four members – a principal investigator and a co-investigator who were both white British women, and two female research fellows of South Asian descent. Maria, whose experiences we reflect upon in this chapter, was one of these research fellows. She was a Pakistani herself and conducted almost all the fieldwork with the Pakistani community. At 29 years of age, she was much younger than the older Pakistani participants, whose ages ranged from 48 to 80 years, but like them, she was herself also a first-generation immigrant in the UK and had been living in the UK independently for several years. All her family was resident in Pakistan and her husband worked in another European country. In terms of her socioeconomic background and origins in Pakistan, she describes herself as being from an urban middle-class background, unlike the majority of the participants who were working class and had migrated from rural areas in Pakistan. Hence, as common with most research, there were both commonalities and differences in the backgrounds and identities of Maria and our participants (see Song & Parker, 1995; Archer, 2002; Yip, 2008; Wray & Bartholomew, 2010). While she shared broadly the same ethnicity with the participants, she described herself as having more of an urban, middle-class, identity – particularly in terms of her gender, and did not strictly adhere to the gendered cultural norms and roles of most of our female participants. It is also important to note here that, unlike the close-knit, same ethnicity, local community networks of most of our participants, Maria had had very little interaction with the local Pakistani community prior to her involvement in the project. Therefore, when she entered 'the field' formally, it was not as a Pakistani person *per se*, but as a Pakistani woman who worked for and represented a White British institution – the university.

Researcher ethnicity and access to gendered community spaces

The use of space among our older Pakistani participants was highly gendered. The older women, unlike most of the older men, not only spent most of their time within the private domestic domain, but also used different spaces to men even within the public spaces of their local Pakistani community. In the local mosques and the Pakistani Community Centre, for example, there were separate rooms designated for women and men, and the two often participated in different, gendered, activities within their local community as well as within the private domestic sphere. This gendered division of space within the community, particularly among its older members, had important implications for gaining access to potential participants, and for Maria's identity in 'the field'. The gendered spatial context of the local ethnic community for instance, powerfully enforced the link between the gender and ethnic identities of not merely its members, but for Maria also, to the extent that these two identities were experienced and perceived as inseparable. This was the case particularly since the cultural scripts and norms for performance (see Goffman, 1959 and 1976; Goddard and Wierzbicka, 2004) of Pakistani ethnic identity vary by gender. For example, linked with the differential, gendered, use of time and space within traditional Pakistani culture, it is both more common and culturally more acceptable for Pakistani men to wear Western clothes. The gendered and age-specific traditional Pakistani cultural norms and discourses in relation to women's roles, bodies and dress, on the other hand, mean that Pakistani women (particularly older ones) are comparatively much less likely to wear Western clothes on a regular basis and would be categorised as being too Western if they did. Hence, given such gendered Pakistani cultural norms and scripts, the authenticity of Maria's Pakistani identity depended to some degree on the extent to which she expressed a specifically female Pakistani identity as defined by the traditional cultural norms of the community.

Since the use of both space and time within the community was gendered, and women (whether young or old) were expected to spend more time within the domestic domain, engaging in activities that centred on their religion and the family (see Drury, 1991; Husband, 1986 cited in Samad, 1998; Hennink et al., 1999), living away from the family in pursuit of a career was not culturally gender-appropriate behaviour. Absence of Maria's family raised curiosity as well as suspicion among some of the community members during the early days of the fieldwork. When initially approached for recruitment, many of the participants and contacts from the local community expressed an interest in learning more about not only Maria herself, but her family members too who were not part of the local community. Participant questions in relation to Maria's family included questions on their whereabouts in Pakistan, her father's and husband's occupations, and one male participant, who seemed surprised to learn that Maria's husband lived in a different country, enquired about his earnings. Such questions were important, and did not merely reflect the nature and topic of our research which explored participant meanings, perceptions and experiences of 'the family', and thus potentially raised participant interest in Maria's family too. In addition, such questions were linked to the gendered use of space within the community in a more general sense, and revealed the importance our participants attached to collective family identities, as opposed to individual and personal identities, particularly for women. Hence, many of the community members would often ask Maria such questions upon meeting her the first time, or finding her unfamiliar presence in their community spaces. This was even when they were unaware that she was a researcher who was carrying out research on the topic of families and caring in South Asian communities. Moreover, on many occasions, such questions were also asked when the participants were approached for recruitment. On these occasions, rather than merely being reciprocal requests for information from the researcher, the questions revealed the extent of participant and community contact concerns about Maria's trustworthiness. Judgements about her identity and trustworthiness often seemed to be linked with judgements about how they perceived her family, as well as in how far they believed her to be prepared to disclose her own personal information.

Personal questions about Maria's family, even when asked with the purpose of checking her credibility and as a means of ensuring that it was 'safe' to participate in the research, often made Maria feel uncomfortable, particularly when such questions were asked by the more powerful and higher status men in the community. This was because such questions made her feel vulnerable, as they exposed that, despite being a Pakistani herself, she did not follow strictly the traditionally prescribed gender norms of the Pakistani culture. Such difficult research encounters challenge the commonly held assumptions within some of the methodological literature, with regards to the balance of power and vulnerability within research relationships when researching under-researched or socially marginalised groups. As pointed out by Russell (1999), in the case of her older participants, such commonly perceived vulnerable and marginalised groups can also be active agents who exercise considerable power over the research process and the research relationship. Russell's observation with regards to her own older participants, and Maria's experience with some of the higher status male participants, brings into question the usefulness of building rapport and trust with participants through an emphasis on one's self-identity and mutual self-disclosure (see Oakley, 1981). While often prescribed as a means of gaining participant trust and equalising the differential power relations when researching marginalised social groups, Maria's experience reveals that in some contexts, rather than merely giving participants equal power, a greater emphasis by the researcher on their self-identity may make the researcher more vulnerable within the research relationship. This had specially been the case within our own fieldwork, where Maria's gender and age interconnected with her Pakistani ethnicity to position her as the less powerful 'other', particularly *vis-à-vis* many of our older male participants.

Given Maria's identity and less powerful status as a young Pakistani woman with fewer of her own social and familial connections with the local Pakistani community, many of our participants – women and men – were able to actively negotiate the terms of their participation within the research. Some reasoned as well as joked with Maria about how they also had the right to have an opportunity to question and interview her in return for all the information they were to disclose to her in their interview. Hence our participants were not in all cases the 'typical' vulnerable group of older ethnic minority people within the specific context of our own fieldwork and 'field' relationships. On the contrary, these participants and community members not merely questioned Maria about her family and herself, but also guided her social behaviour in their own community spaces, bringing it in line with their own cultural norms. In doing so, they seemed to exercise their control over the research relationship by setting the terms for Maria's acceptance within their community spaces.

Maria's initially perceived status within the community as neither a complete 'insider' nor an 'outsider', and particularly the absence of her family from the local scene, nevertheless also raised suspicion among community members and participants with regard to her sincerity towards the community in carrying out the research *vis-à-vis* her own career aspirations in a White public domain. A certain amount of self-disclosure was thus crucial for gaining initial access to and acceptance within these community spaces – both private and public. Since not having any family living close by is culturally unusual for a young Pakistani woman, and seemed to clearly position Maria initially as an 'outsider' within the community, she actively needed to express and negotiate her 'insider' status within the community. As part of her negotiation, Maria often justified the absence of her family, and the reasons why she chose to live and work in a different country to her husband, in terms that were more culturally acceptable to the participants – such as the difficulties and the long procedures involved in obtaining visas for work and settlement. While Maria's own personal identity and her own cultural values and ideologies on gender may be seen to conflict with those of the participants, it was important for her to downplay these cultural differences in order to build a good rapport with the participants, and to be more open to learning about and gaining an insight into their lives, experiences and values, meanings and perspectives.

The requirement to build up rapport with participants as a means of gaining participant trust and as a tool for easy access to 'the field', has recently been much debated in terms of its utility and ethics (Duncombe & Jessop, 2002; Gaglio et al., 2006; Knowles, 2006; Smyth & Mitchell, 2008). While some researchers and scholars have seen it as useful (e.g. Oakley, 1981; Ellingson, 1998), others have expressed their discomfort with the potential ethical issues it raises (see Duncombe & Jessop, 2002). Within the context of our own fieldwork, however, we deemed it important to recognise the multiplicity of Maria's identities. Despite the difference in how she negotiated her own Pakistani gender identity in relation to that of most of the female participants, she still shared with them her subjective identity of being a Pakistani woman, and a similar (if not the same) experience of being a first-generation migrant in the UK. The gendered division of space within the community, nevertheless, meant that in practice it was not enough for her to merely be a Pakistani woman. An unobtrusive access to and entry in these gendered spaces required her to embrace and perform the clearly defined and visible gendered identity of a Pakistani woman (see Zubair et al., in press). In this respect, rather than being a matter for deliberation and choice, the particular sociocultural context of our participants imposed this gendered identity on her from the time she entered 'the field'.

In discussing Maria's identity in 'the field', it is also important to recognise the context of her initial entry as she tried to gain access to and engage with the local community. The start of her fieldwork with this community coincided with the month of Ramadan, for Muslims, a holy month of fasting and praying. The increased focus on religion, spirituality and the family during this month, and the limitations of social activities other than prayers amongst most of the older members of the community, meant that the only space where Maria was able to access potential participants was either at the Pakistani Community Centre or at the local mosques during prayer time. Her attendance and participation in these prayers night after night, dressed in the traditional female Pakistani Muslim clothes (see Figure 4.1), raised cultural expectations among our participants for continued gender appropriate embodied behaviour and use of space (see Zubair et al., in press) and influenced the character of her interactions with participants. In particular, it led to the formation of gendered research relationships between her and the female and male participants. Entering male Pakistani spaces became much more difficult than before, although this did not seem to affect the recruitment of male participants, who appeared much more willing to participate in the research than the women. On the other hand, the closer and more informal relationships formed with the women put much more demands on Maria's time, as the women expected her to spend more time with them during her visits to the community centres, the mosques and their private domestic spaces. Despite this, the women in the community remained less willing to participate in the research. This unusual variation between the female and the male willingness to participate was linked with the issue of gatekeepers, to which we now turn.

Gatekeepers and barriers to women's recruitment/ participation

We encountered multiple socio-cultural barriers to the recruitment of female participants for our research. Many of the older women in the community had had little or no schooling and thus often lacked the relevant literacy skills, even within their own spoken languages, for reading and making sense of the information we provided to them in the information leaflets. There were similar issues with the reading and signing of the consent forms (Zubair et al., 2009). In addition to the issue of literacy, the gendered use of time and space meant that the women, particularly the older ones, had more of a domestic role within the community, and

Figure 4.1

were not used to interacting and communicating with formal official agencies without intermediaries. The recruitment of the female participants was further complicated by a general mistrust of White officialdom among many community members. As we represented the School of Health and Social Care at our university, an identity we had used in our information leaflets to advertise ourselves to the local Pakistani community, we were more often than not perceived within the community as linked with social services. This became increasingly obvious to us during the fieldwork when, on different occasions, some of the older women referred to Maria as a social worker. Many of these women declined participation in the research explaining they were unable to make the time for it, or were not in the right health to do so. However, different community sources indicated to us that the actual reason for refusal by many women was their fear of disclosing by mistake any information that may later be used as a 'proof' against them (see Levkoff & Sanchez, 2003).

Accessing many 'hard to reach' populations is problematic, and often such difficulties are exacerbated by the role of 'gatekeepers'. In our study too, the differential recruitment of women and men reflected the role of gatekeepers – the male members of the women's family and the women's children. Boneham (2002) also recognises how older South Asian women's voices and experiences often remain hidden within research because male community leaders and family members become important intermediaries, who speak on behalf of their female family members. Boneham sees the issue as arising from the linguistic and cultural differences of the researchers from these older ethnic minority women who, unlike their male counterparts, are less fluent in the English language. Seeing the barrier to the recruitment and participation as lying in the linguistic and cultural differences between the researchers and these older ethnic minority participants, Boneham (2002) proposes that women researchers of the same cultural background as the older South Asian women should be included in the research process. Our experience of fieldwork and recruitment with the Pakistani community, however, suggests that the issue is more complex. A simple 'matching' of the ethnicity and gender of the researcher with that of the participants does not, of itself, ensure success in recruitment. This is because, in addition to the socio-cultural barriers, the issue of recruitment of ethnic minorities in research is further complicated by the issue of trust (see Kreiger et al., 2001; Barata et al., 2006; Gallagher-Thompson et al., 2006; Feldman et al., 2008) and representation (see Levkoff & Sanchez, 2003; Yancey et al., 2006). Hence a researcher of the same South Asian ethnicity and gender who has the appropriate linguistic and cultural skills, but is representing a White institution and differs from the female participants in terms of other social characteristics, may not be much more successful in gaining access and recruitment than a researcher who is an ethnic 'outsider'.

Despite being an ethnic 'insider' but differing from many of our participants along other important, interconnected, dimensions of identity (for example, her social class, urban Pakistani background and professional affiliation), Maria experienced the powerful role of the gatekeepers throughout her fieldwork with the women (and some of the lower status men) in the community. This was the case even when she had gained direct access to the women's private as well as public spaces in the community. On many occasions during the fieldwork, for example, women who seemed to have happily agreed to participate changed their minds after having discussed it with their families. Some of these participants and contacts hinted that such changes of mind related to a combination of factors including: a genuine and increased suspicion of the purposes of the research after discussions with family members – particularly with the husbands and adult children; a greater trust in the judgements of the more 'knowledgeable' family members who were not in favour of participation; but also a sense of family loyalty whereby it was not deemed right to participate in something which the family had advised against. Many other women who seemed willing and happy to participate suggested to Maria that they could participate if she met their families and explained the purposes of the project to their husbands or adult children. Hence, while the ethics and usefulness of involving gatekeepers within the research process remains a topic for debate (see Miller, 1998; Gallagher-Thompson et al., 2006; Feldman et al., 2008; Sheikh et al., 2009), within our research the gatekeeper's role involved not merely providing referrals and assurances to potential participants, but in addition, to remain a key figure during the fieldwork, without whose approval the older female participants were less likely to participate.

The important role of the gatekeepers continued even in those instances where Maria had gained access to participants directly herself. The particular family dynamics of the participants meant that even consent for participation had to be sought not merely from the female participants themselves, but we also required the consent of the important gatekeepers within the family (see Levkoff & Sanchez, 2003) in the form of a tacit approval for their female family members' participation. As described above, participants and their family members often asked Maria questions about herself and, in some cases, even verified the information she gave relating to herself and her family by checking it informally with other participants and mutual contacts in the community who had known Maria previously. Fewer questions, on the other hand, were asked about the research project itself. Many of the participants and their family members did not fully read the information leaflets and the consent forms, reasoning that they trusted what Maria had already explained to them. While their propensity to not read the information leaflets we provided to them relates to the mainly oral character of the indigenous rural Pakistani culture (Zubair et al., 2009), this also reveals in part the importance they attached to the perceived identity and trustworthiness of Maria.

Maria's embodied identity in 'the field'

We have illustrated above how the recruitment of the older Pakistani women was not easier for Maria despite being a woman and a Pakistani herself. The multiple and cross-cutting nature of social identities meant that, rather than being a complete 'insider', Maria was faced with various multi-layered and complex insider/outsider boundaries within 'the field' (see Archer, 2002; Yip, 2008; Wray & Bartholomew, 2010; Zubair et al., in press). As we argue in another paper (see Zubair et al., in press), within this context, Maria's rapport with the participants, and their trust in her, developed on the basis of her embodied and visible Pakistani identity, for example through her dress and culturally appropriate bodily participation in community public and private spaces.

We have described above how Maria's initial entry in 'the field' took place during Ramadan, when she attended night prayers at the local Pakistani community centre and the local mosque. From the moment of this initial fieldwork phase, Maria's dress and bodily behaviour remained important markers of her identity in 'the field' on the basis of which, most participants related to her. Differences in dress and bodily behaviour were often met with different reactions by the participants. Most appeared more comfortable and friendly when Maria's dress and body were more in line with their own specific Pakistani cultural norms. Hence, even though Maria chose to wear the traditional Pakistani dress rather than Western dress during most of the fieldwork (see Figure 4.2), given the internal cultural variations within the community, she often adjusted her Pakistani dressing style in accordance with more specific Pakistani cultural norms and expectations (see Figures 4.1–4.3). Such adaptations in dress were important for gaining community members' approval. On many occasions, for example, upon finding that Maria's dress did not fully meet their own specific cultural norms, both the female and male participants commented on Maria's attire – at times, encouraging her to dress 'appropriately' to their norms, and in a couple of instances expressing disapproval for her particular use of dress. On one occasion, for example, an older male participant encouraged Maria to cover her head with a scarf at all times (as opposed to only when offering prayers) explaining to her how, within the Pakistani community 'a woman looks more decent, respectable and beautiful when she has her head covered'. On two more occasions, a

Figure 4.2

female and a male participant who similarly liked the younger women to wear headscarves, offered to buy Maria these as gifts so that she would wear them more regularly than they thought she did. While Maria politely declined these offers of gifts, explaining to the participants that she would be able to buy these locally herself, she felt a

Figure 4.3

greater need to adhere to these participant specific expectations and norms. These participant attempts to socialise Maria into a culturally appropriate use of dress revealed the tensions between her 'insider' and 'outsider' status *vis-à-vis* these older Pakistani participants. On the one hand, they seemed to perceive her as a younger member of the

Pakistani community but, on the other, they were also conscious of her cultural and social differences, despite her Pakistani ethnicity. Their concerns around Maria's dress furthermore revealed the significance they attached to her embodied female Pakistani identity.

We have discussed earlier how our participant and contact questions about Maria's family were often a check on her trustworthiness, as well as being based on their curiosity to learn more about her. Maria found her perceived cultural respectability – as judged within this community on the basis of both her family identity and her individual social behaviour – and her cultural authenticity were often important for gaining referrals and a snowball sample. Her contacts within the community, particularly the women, when introducing her to potential participants for recruitment purposes described her, on the basis of her Pakistani ethnicity and her family identity. Contacts and participants, for example, would often describe her to others as '*Apni ladki/kudi* (Our own girl) from Multan'. One female participant and key contact consistently introduced Maria to many other potential participants by informing them that Maria was not only working at the university (and thus educated), but that she was also married. The latter attribute was clearly perceived and presented by this contact as an achievement, and as something that added to Maria's cultural authenticity, and more so, to her respectability. In addition to her family identity, judgements about Maria and her cultural authenticity were also linked to her embodied identity and presence within the community. Hence, throughout the fieldwork, female participants and contacts often also referred to Maria's embodied religious and ethnic identity as they told friends and families how Maria had been a regular attendee throughout Ramadan prayers. All these aspects of Maria's identity endorsed her shared ethnicity with community members, downplayed the white officialdom that may be associated with somebody working for a White institution, and made her appear potentially more trustworthy.

Concluding comments: Implications for fieldwork with ethnic minority older people

The issues of access and recruitment in qualitative research with older people belonging to ethnic minority communities are complex as these relate not merely to the potential cultural and linguistic differences between researchers and participants but, as in the case of qualitative research in general, these also implicate the researcher's identity within the process. In this chapter, we have attempted to illustrate some of these complexities through examples of Maria's experiences of fieldwork with our older

Pakistani Muslim participants. In particular, we have documented the arbitrary nature of the insider/outsider distinction and reflected upon how a researcher may be perceived both as an 'insider' and as an 'outsider' depending on the specific context of their interaction with participants. It is thus important for qualitative researchers who research older people from ethnic minority populations to recognise that the simplistic binaries of insider/outsider based on a singular identity or researcher characteristic, masks the actual dynamic, fluid and contextual nature of most research relationships as they unfold and are performed within 'the field'.

Within our fieldwork, Maria's ethnic identity interlinked with her gender, age and other social identities to influence the particular character of her interactions with participants. Her experiences in 'the field' as a Pakistani researcher undertaking research with the Pakistani community in Britain illuminate well the arbitrariness of the insider/outsider distinction. These experiences reveal that, while a researcher's success in relation to access and recruitment of participants remains linked to issues of trust, this trust may not be produced simply through a 'matching' of researcher and participant characteristics. We have shown how, for Maria, *being* a Pakistani *per se* neither ensured easier acceptance within gendered community spaces, nor did it help considerably with recruitment. Successful access and recruitment, on the contrary, involved the active negotiation of an 'insider' status and a visible, embodied, performance of gendered Pakistani ethnicity. Hence, ethnic or other forms of 'matching' are likely to be helpful in building trust and rapport only in so far as the researcher can also actively express and negotiate their cultural authenticity along other important and contextually relevant, interconnected, dimensions of their personal, social and familial identities. Assuming complete similarity (or dissimilarity) with one's participants, is often misleading and problematic within research. It was, for example, largely through Maria's recognition of how participants differed from her in terms of their gender ideologies and family norms and dynamics, that she was able to accommodate these differences within the research process and negotiate her own cultural authenticity.

Annotated reading list

Butler, J. (1993) *Bodies that Matter: On the Discursive Limits of 'Sex'* (London: Routledge).

In this influential work, Butler uses the concept of performativity to lay open the socially constructed nature of gender and gender identities. For her, gender is not about being something – i.e. an expression of what one is, but it's something that one does. Butler's ideas have been extended and applied to other forms of social identities such as ethnicity, social class and age which are seen to be performed in particular ways to negotiate inclusion into (or social distance from) certain social groups. While the book itself focuses heavily on gender and performance theory, within our own work, these theoretical underpinnings have been important in informing greater reflexivity in terms of researcher-participant interactions and representations of their own identities.

Ellingson, L.L. (2006) 'Embodied knowledge: Writing researchers' bodies into qualitative health research', *Qualitative Health Research*, 16 (2) 298–310.

Ellingson stresses the need for qualitative researchers to be reflexive with regards to their bodily experiences and interactions in writing up their fieldwork accounts. Drawing upon examples of her own embodied experiences during fieldwork with a geriatric oncology team and their patients, she reveals the usefulness of embodied fieldwork accounts and illustrates ways in which researchers can make their research writing embodied.

Goffman, E. (1959) *The Presentation of Self in Everyday Life* (New York: Anchor Books).

Goffman, in this influential work, compares much of everyday life, micro-level, social interaction and social life to a theatrical production. He views individuals and social groups as putting on performances in their daily life social interactions akin to actors on a theatrical stage. Through following the appropriate cultural scripts in their performance during a given social interaction – including, for example, the use of appropriate scripted dialogues, gestures, rituals, props, costumes, emotions and so on – actors try to convey to an audience a particular presentation of their own self.

Gunaratnam, Y. (2003) *Researching 'Race' and Ethnicity: Methods, Knowledge and Power* (London: Sage).

This book offers some useful reading on the challenges involved in researching race and ethnicity qualitatively. Using case study examples, Gunaratnam addresses the issue of researching across social differences and also illustrates how race and ethnicity is produced, negotiated and resisted in qualitative research encounters.

Oakley, A. (1981) 'Interviewing women: A contradiction in terms' in H. Roberts (ed.) *Doing Feminist Research*, pp. 30–61 (London: Routledge and Kegan Paul).

In this well-renowned work, Oakley has advocated the use of self-identity and mutual self-disclosure by researchers as a means of building rapport and intimate, non-hierarchical, relationships with their participants. While this emphasis on building rapport and intimate research relationships with participants has recently come under much scrutiny, it remains an important stance within most qualitative research on under-researched and socially marginalised groups.

Okely, J. (2007) 'Fieldwork embodied' in C. Shilling (ed.) *Embodying Sociology: Retrospect, Progress and Prospects*, pp. 65–79 (Oxford: Blackwell).

Okely describes the important role that researchers' bodies – including their bodily appearances, bodily adaptations, bodily participation and bodily interactions, occupy within fieldwork. Using specific examples of researchers' embodied experiences and interactions during fieldwork, she demonstrates how fieldwork is a highly embodied activity.

Acknowledgements

This study is funded by grant reference RES-352-25-0009A as part of the ESRC *New Dynamics of Ageing* Programme directed by Professor Alan Walker.

We wish to formally acknowledge the work of Dr Subrata Saha on the project between October 2007 and February 2010. We are grateful for the support and participation of the local communities and to all those who participated in the study. We are also grateful to Professor Julia Twigg for her help in identifying some useful literature on the performativity of identities.

References

Archer, L. (2002) '"It's easier that you're a girl and that you're Asian": Interactions of "race" and gender between researchers and participants', *Feminist Review*, 72 (1) 108–32.

Barata, P.C., Gucciardi, E., Ahmad, F. & Stewart, D.E. (2006) 'Cross-cultural perspectives on research participation and informed consent', *Social Science and Medicine*, 62 (2) 479–90.

Bhavnani, K.K. (1991) 'Tracing the contours: Feminist research and feminist objectivity', *Women's Studies International Forum*, 16 (2) 95–104.

Bhopal, K. (2001) 'Researching South Asian women: Issues of sameness and difference in the research process', *Journal of Gender Studies*, 10 (3) 279–86.

Boneham, M. (2002) 'Researching ageing in different cultures' in A. Jamieson & C.R. Victor (eds) *Researching Ageing and Later Life: The Practice of Social Gerontology*, pp. 197–210 (Buckingham and Philadelphia: Open University Press).

Burholt, V. & Wenger, G.C. (2003) *Families and Migration: Older People from South Asia* (University of Wales, Bangor: Centre for Social Policy Research and Development).

Drury, B. (1991) 'Sikh girls and the maintenance of an ethnic culture', *New Community*, 17 (3) 387–99.

Duncombe, J. & Jessop, J. (2002) '"Doing rapport" and the ethics of "faking friendship"' in M. Mauthner, M. Birch, J. Jessop & T. Miller (eds) *Ethics in Qualitative Research*, pp. 107–22 (London, Thousand Oaks and New Delhi: Sage Publications).

Edwards, R. (1990) 'Connecting method and epistemology: A white woman interviewing black women', *Women's Studies International Forum*, 13 (5) 477–90.

Ellingson, L.L. (1998) '"Then you know how I feel": Empathy, identification, and reflexivity in fieldwork', *Qualitative Inquiry*, 4 (4) 492–514.

Feldman, S., Radermacher, H., Browning, C., Bird, S. & Thomas, S. (2008) 'Challenges of recruitment and retention of older people from culturally diverse communities in research', *Ageing and Society*, 28 (4) 473–93.

Gaglio, B., Nelson, C.C. & King, D. (2006) 'The role of rapport: Lessons learned from conducting research in a primary care setting', *Qualitative Health Research*, 16 (5) 723–34.

Gallagher-Thompson, D., Rabinowitz, Y., Tang, P.C., Tse, C., Kwo, E., Hsu, S., Wang, P.C., Leung, L., Tong, H.Q. & Thompson, L.W. (2006) 'Recruiting Chinese Americans for dementia caregiver intervention research: Suggestions for success', *American Journal of Geriatric Psychiatry*, 14 (8) 676–83.

Goddard, C. & Wierzbicka, A. (2004) 'Cultural scripts: What are they and what are they good for?', *Intercultural Pragmatics*, 1 (2) 153–66.

Goffman, E. (1959) *The Presentation of Self in Everyday Life* (New York: Anchor Books).

Goffman, E. (1976) 'Gender display', *Studies in Anthropology of Visual Communications*, 3, 69–77.

Hanna, L., Hunt, S. & Bhopal, R. (2008) 'Insights from research on cross-cultural validation of health-related questionnaires: The role of bilingual project workers and lay participants', *Current Sociology*, 56 (1) 115–31.

Harding, S. & Balarajan, R. (2001) 'Longitudinal study of socio-economic differences in mortality among South Asian and West Indian migrants', *Ethnicity and Health*, 6 (2) 121–8.

Hennink, M., Diamond, I. & Cooper, P. (1999) 'Young Asian women and relationships: Traditional or transitional?', *Ethnic and Racial Studies*, 22 (5) 867–91.

Katbamna, S., Bhakta, R., Ahmad, W., Baker, R. & Parker, G. (2002) 'Supporting South Asian carers and those they care for: The role of the primary health care team', *British Journal of General Practice*, 52, 300–5.

Knowles, C. (2006) 'Handling your baggage in the field: Reflections on research relationships', *International Journal of Social Research Methodology*, 9 (5) 393–404.

Kreiger, N., Ashbury, F., Cotterchio, M. & Macey, J. (2001) 'A qualitative study of subject recruitment for familial cancer research', *AEP*, 11 (4) 219–24.

Lee, A. (2008) 'Finding the way to the end of the rainbow: A researcher's insight investigating British older gay men's lives', *Sociological Research Online*, 13 (1) 6.

Levkoff, S. & Sanchez, H. (2003) 'Lessons learned about minority recruitment and retention from the centers on minority aging and health promotion', *The Gerontologist*, 43 (1) 18–26.

Lloyd, C.E., Johnson, M.R.D., Mughal, S., Sturt, J.A., Collins, G.S., Roy, T., Bibi, R. & Barnett, A.H. (2008) 'Securing recruitment and obtaining informed consent in minority ethnic groups in the UK', *BMC Health Services Research*, 8 (68).

McLean, C.A. & Campbell, C.M. (2003) 'Locating research informants in a multi-ethnic community: Ethnic identities, social networks and recruitment methods', *Ethnicity and Health*, 8 (1) 41–61.

Merrell, J., Kinsella, F., Murphy, F., Philpin, S. & Ali, A. (2006) 'Accessibility and equity of health and social care services: Exploring the views and experiences of Bangladeshi carers in South Wales, UK', *Health and Social Care in the Community*, 14 (3) 197–205.

Miller, T. (1998) 'Shifting layers of professional, lay and personal narratives: Longitudinal childbirth research' in J. Ribbens & R. Edwards (eds) *Feminist Dilemmas in Qualitative Research: Public Knowledge and Private Lives*, pp. 58–71 (London, Thousand Oaks and New Delhi: Sage Publications).

Moreno-John, G., Gachie, A., Fleming, C.M., Napoles-Springer, A., Mutran, E., Manson, S.M. & Perez-Stable, E.J. (2004) 'Ethnic minority older adults participating in clinical research: Developing trust', *Journal of Aging and Health*, 16 (5) 93S–123S.

Nazroo, J. (2006) 'Ethnicity and old age' in J. Vincent, C. Phillipson & M. Downs (eds) *The Futures of Old Age*, pp. 62–72 (London: Sage Publications).

Nazroo, J., Bajekal, M., Blane, D. & Grewal, I. (2004) 'Ethnic inequalities' in A. Walker & C. Hennessey (eds) *Growing Older: Quality of Life in Old Age*, pp. 35–59 (Maidenhead: Open University Press).

Oakley, A. (1981) 'Interviewing women: A contradiction in terms' in H. Roberts (ed.) *Doing Feminist Research*, pp. 30–61 (London: Routledge and Kegan Paul).

Phillipson, C., Ahmed, N. & Latimer, J. (2003) *Women in Transition: A Study of the Experiences of Bangladeshi Women Living in Tower Hamlets* (Bristol: The Policy Press).

Qureshi, T. (1998) *Living in Britain – Growing Old in Britain: A Study of Bangladeshi Elders in London* (London: Centre for Policy on Ageing).

Ramji, H. (2008) 'Exploring commonality and difference in in-depth interviewing: A case-study of researching British Asian women', *The British Journal of Sociology*, 59 (1) 99–116.

Russell, C. (1999) 'Interviewing vulnerable old people: Ethical and methodological implications of imagining our subjects', *Journal of Aging Studies*, 13 (4) 403–17.

Samad, Y. (1998) 'Media and Muslim identity: Intersections of generation and gender', *Innovation: The European Journal of Social Sciences*, 11 (1) 425–38.

Sheikh, A., Halani, L., Bhopal, R., Netuveli, G., Partridge, M.R., Car, J., Griffiths, C. & Levy, M. (2009) 'Facilitating the recruitment of minority ethnic people into research: Qualitative case study of South Asians and asthma', *PLoS Medicine*, 6 (10).

Skeggs, B. (1997) *Formations of Class and Gender* (London: Sage Publications).

Smyth, L. & Mitchell, C. (2008) 'Researching conservative groups: Rapport and understanding across moral and political boundaries', *International Journal of Social Research Methodology*, 11 (5) 441–52.

Song, M. & Parker, D. (1995) 'Commonality, difference and the dynamics of disclosure in in-depth interviewing', *Sociology*, 29 (2) 241–56.

Twine, F.W. (2000) 'Racial ideologies and racial methodologies' in F.W. Twine & J.W. Warren (eds) *Racing Research, Researching Race: Methodological Dilemmas in Critical Race Studies*, pp. 1–34 (New York: New York University Press).

Victor, C.R., Martin, W. & Zubair, M. (2012) 'Families and caring amongst older people in South Asian communities in the UK: A pilot study', *European Journal of Social Work*, 15 (1) 81–96.

Vincent, J., Phillipson, C. & Downs, M. (2006) *The Futures of Old Age* (London: Sage Publications).

Wray, S. & Bartholomew, M. (2010) 'Some reflections on outsider and insider identities in ethnic and migrant qualitative research', *Migration Letters*, 7 (1) 7–16.

Yancey, A.K., Ortega, A.N. and Kumanyika, S.K. (2006) 'Effective recruitment and retention of minority research participants', *Annual Review of Public Health*, 27(1) 1–28.

Yip, A.K.T. (2008) 'Researching lesbian, gay, and bisexual Christians and Muslims: Some thematic reflections', *Sociological Research Online*, 13 (1).

Zubair, M., Martin, W. & Victor, C. (2009) 'Exploring gender, age, time and space when researching older people living in Pakistani Muslim communities in the UK: Reflections from the "field"', paper presented at *British Society of Gerontology (BSG) 38th Annual Conference*, Bristol, UK.

Zubair, M., Martin, W. & Victor, C. (2010) 'Researching ethnicity: Critical reflections on conducting qualitative research with people growing older in Pakistani Muslim communities in the UK', *Generations Review: Journal of the British Society of Gerontology*, 20 (1), http://www.britishgerontology.org/DB/gr-editions-2/generations-review/researching-ethnicity-critical-reflections-on-cond.html

Zubair, M., Martin, W. & Victor, C. (in press) 'Embodying gender, age, ethnicity and power in "the field": Reflections on dress and the presentation of the self in research with older Pakistani Muslims', *Sociological Research Online*.

5
Piecing Together Experiences of Older People with Intellectual Disability

Christine Bigby

People with intellectual disability are one of the new non-traditional groups who are beginning to form part of the ageing population (see Chapter 3). It is only in the last 40 years that significant numbers with intellectual disability have survived into adulthood and later life. For example, the average life span of people with Downs Syndrome has increased from 30–35 years to 55–60 years since the 1960s (Torr et al., 2010). As well as the new experiences of ageing, their increased life expectancy has led to hitherto unexplored phenomena such as 'ageing carers', interdependent households comprising aged parents with middle-aged sons and daughters with intellectual disability, and new types of relationships such as older siblings with and without an intellectual disability. Such developments should be a cause for celebration, which exercises our imagination to find ways to maximise quality of life as people age, and to uncover the expectations and experiences about this new part of the lifecourse. Too often however, they are cast in a negative light, with attention turned to the problems caused for already overstretched service systems.

Increased longevity does pose challenges unique to people with intellectual disability. Knowledge about earlier parts of the lifecourse suggests they age from a quite different and often very disadvantageous position compared to many other population groups. People with intellectual disability are much more likely to lack the social and monetary capital accumulated during adulthood, and which provide the foundations for choice, consumer power, participation and good health in later life (Marmot & Wilkinson, 1999). For example, few will have held full-time paid employment, accumulated superannuation, bought their own home, married or had children (Bigby, 2004). Most will have lived with their parents until well into middle age. For example, the ratio of people with intellectual disability living in the family home compared to supported accommodation changes from 70:30 to 30:70 after the age of 55 years (Emerson et al., 2001). People with intellectual disability are more likely than others who embark on the ageing process to have led inactive lives, been overweight, experienced poverty, and been subjected to abuse (Emerson et al., 2005). Despite 30 years of deinstitutionalisation people with intellectual disability continue to occupy a 'distinct social space' with small, dense social networks made up only of paid staff, family members and other people with intellectual disability (Clement & Bigby, 2009). As Reinders (2002) suggests policy can create the right to occupy formal roles such as citizen, employee and homeowner, but cannot mandate 'civic friendship' in the form of convivial social relationships.

The death of their parents and lifelong primary caregivers is the crisis of mid-life for people with intellectual disability who remain at home, which is also too often associated with disruption to their lives and a move to some form of supported accommodation. Thus, the place where many people with intellectual disability will hope to 'age in place' will be a disability service, most likely a group home. Yet aside from questions about such living situations being more restrictive than necessary and curtailing independence, long-term residence in such 'homes' is jeopardised by the difficulties formal services have in adapting to changed support needs. The disproportionate number of younger older people with intellectual disability who are prematurely resident in aged care facilities, and their longer periods of residence compared to other residents, attest to this (Bigby et al., 2008; Thompson et al., 2004).

The rigid age-based criteria that regulate access to specialist age-related health and social services create formidable obstacles to gaining appropriate support and can add a further layer of disadvantage. Many people with intellectual disability do not conform to standard ageing trajectories or associated chronological definitions of being old. For example, people with Downs Syndrome are susceptible to early onset Alzheimer's disease in their 40s and 50s, and people with other genetic conditions such as Fragile X or multiple, complex needs experience premature often complicated ageing processes. They require access to age-related diagnostic services, memory and falls clinics in their 50s rather than after the age of 65. Given all of these factors it is not surprising that large-scale epidemiological and health-related studies show the relatively poor health status and high number of unidentified and untreated medical conditions of older people with intellectual disability (Haveman et al., 2010).

Significantly, over the past decade, at the broadest level, ideas and policy about both ageing and disability have moved in unison from devaluing policies that fostered dependency, segregation, passivity and disengagement, to the current more positive visions such as 'active ageing' (WHO, 2002) and 'full and effective participation and inclusion in society' (United Nations, 2006), founded on rights, autonomy and older disabled people's participation. Disability policy is now firmly located in a rights paradigm that suggests that older people with intellectual disability should expect comparable opportunities to the general population for a 'good old age'.

Despite these overarching frameworks and the urging of researchers, that stretch back to the mid-1980s, a policy vacuum remains in western countries whereby aged care, disability and health services have few imperatives to provide services adapted to the needs of ageing people with intellectual disability (Bigby, 2010). Yet this group is ageing from an already disadvantaged position, and the very nature of their intellectual impairment means they are likely to require more support than others to navigate service systems, advocate for their own needs, exercise choice about their lives, participate in meaningful activities and maintain relationships.

Bigby and Knox (2009) suggest that as people with intellectual disability age, their lives are fragmented between formal and informal worlds, with no one having a sense of the complete picture. Very little is known about the lives of older people with intellectual disability from their own standpoint. Intellectual impairment means people have difficulty reflecting and talking about their own lives. People with intellectual disability seldom write their autobiographies, and their status in society means they rarely become sufficiently well known to attract the attention of biographers. A notable exception is the work of the Learning Disability History Research group of the Open University, led by Dorothy Atkinson, which has supported people to research and publish their life stories (http://www.open.ac.uk/hsc/ldsite/research_grp.html). Research can play a significant role in uncovering the expectations and the lived experiences of ageing people with intellectual disability and their families, as well as the mechanisms and values that underpin service and policy responses to them. Research, the evidence and ammunition for advocacy that flows from it, as well as the demonstration of what is possible by using methods such as action research is one strategy to ensure ageing is a cause for celebration and not a curse for people with intellectual disability.

Little account has been taken of this minority group by gerontologists or mainstream social researchers. Communication difficulties, robust gatekeeping, reliance on others as proxy informants, as well as a failure to differentiate sufficiently among people with disabilities mean people with intellectual disability have not figured as respondents to large-scale population surveys of older people. The inclusion of an intellectual disability supplement to the Irish Longitudinal Study on Aging (TILDA) is a remarkable exception, demonstrating both the possibilities and challenges of including people with intellectual disability in large-scale surveys. Whilst intellectual disability-specific research about ageing issues has grown exponentially in the past 20 years, only a small fraction has used qualitative methods to explore directly or in any depth the experiences of ageing from the stance of people themselves. Some research has sought the perspective of older parents, particularly about their anxieties for the future care of their middle-aged son or daughter, and the experiences of older carers have been particularly well represented by advocacy groups in Australia (www.carersaustralia.com.au/), and in the work of organisations like the Foundation for People with Learning Disabilities in the UK (http://www.learningdisabilities.org.uk/). In many ways however, this has served to give ownership of issues associated with ageing to parents rather than to people themselves and has problematised ageing of people with intellectual disability.

Factors, such as hidden populations, ethical recruitment practices, issues of consent and resources have created formidable obstacles to the inclusion of the perspectives of older people with intellectual disability in research. This chapter explores some of the issues that confront researchers who are committed to ensuring the voices of people with intellectual disability are heard and their interests represented in research. Issues revolve primarily around how researchers can access participants from a diversity of living circumstances and not just service users, and which data collection strategies might best capture the views of people with intellectual disability, or of others who purport to represent them. Finally I consider some of the distinctive challenges of occupying multiple roles in intervention or action research.

Locating a 'hidden' population

Much of the research on ageing parents and older people with intellectual disability has recruited participants from among the easily identifiable group of disability service users by canvassing agencies or the use of local or national registers of people with intellectual disability. Such strategies effectively exclude those out of touch with services. Early research from the United States of America (USA) and the United Kingdom (UK) estimates that 60–75 per cent of the population of older people with intellectual disability are unknown to the disability service system (Hogg et al., 1988; Jacobson et al., 1985; Krauss & Seltzer, 1986). A later Australian study found a smaller but still significant figure of 28 per cent for the hidden group (Bigby, 1995). They are thought to comprise older adults living with parents who have rejected institutionalisation or became alienated from services or people with mild intellectual impairment who lived independently after the first waves of deinstitutionalisation. Researchers have used various strategies to find hidden groups through contact with their local communities. For example, in the US Janicki and his colleagues (Janicki et al., 1998) report that area agencies on ageing successfully use outreach methods such as workshops, public service announcements and appearances, stories in the local press, contacts with clergy, police, postal workers and pharmacists, and postings of notices in places such as markets, senior centres, community facilities, public offices, and clinics. In Australia, Llewellyn et al. (2010) located older carers through local domiciliary and other mainstream older person's day, residential or health assessment services, general practitioners and support groups. In my 1994 study I used similar strategies to find older people with intellectual disability, and although this was prior to the email revolution, I would still contend that regular phone calls and face-to-face contact are far more useful in prompting action than written communication. The continuing confusion however, between intellectual disability and mental illness can inflate the hidden population, and some form of initial screening of those identified is necessary.

In research about the history of a self-advocacy organisation, we worked with members of the committee using their old photos and documents to identify past supporters, associates and members. We then searched phone directories and the electoral roll, and used a networking approach to hunt down contact details which we used to invite people to a self-advocacy reunion. This proved not only to be a very effective strategy to recruit to the research out-of-touch members and others who had played a role in the organisation's past, but also created the opportunity for social connections to be re-established, and a celebration of the organisations achievements. This illustrates also the multiple purposes and outcomes of collaborative research with people with intellectual disability (Bigby et al., 2010).

The important message from all these strategies is that few people are well hidden, and most are known to someone or a service system other than disability. It takes ingenuity and time to locate them, but such effort adds to the research quality, ensuring greater diversity among participants. The low literacy rates of people with intellectual disability and their need for support to access and understand information may mean that more conventional methods of reaching out to recruit hidden groups in a population, such as advertising, use of internet lists, or social networking technologies may not be as successful.

Getting past the gatekeepers

Older people who use disability services may be an easily identified population but are nevertheless difficult to recruit as research participants. Risk adverse criteria of institutional Human Research Ethics committees increasingly require formal approval from every organisation asked to assist in recruiting service users. This, together with a failure of organisations to reciprocate each other's approvals, means multiple ethics applications are required, too often on differing forms. For example, to conduct the intellectual disability supplement to TILDA, researchers had to apply to more than 20 different committees before the recruitment process could commence. Ethical practice, aimed at avoiding participant coercion requires an intermediary to act as a conduit for invitations, and information about research projects from researcher to potential participants. This is a different situation from simply asking organisations to address and post letters to clients. Receiving information is not enough, and service users with intellectual disability, no matter how simple the English, are likely to need support to understand and make decisions about acting on an invitation to participate. This means researchers are reliant on the goodwill of already overstretched staff to find the time to distribute and explain projects to service users. At this point staff make implicit judgements about an individual's willingness or interest to participate, by determining to whom material is distributed, and with how much enthusiasm or support. Staff will also make decisions about peoples' capacity to consent, often bypassing service users and giving information directly to family members, who then make decisions about participation. Just ensuring information about research is distributed to users across large organisations can be a lengthy and unpredictable process, with many pitfalls, especially where services are dispersed and staff work shifts. It requires dogged perseverance and time, but to whom information has been distributed, in what manner, and what screening occurs of antagonistic clients or outspoken families, usually remains unknown.

Gaining consent

A considerable literature exists on decision-making by people with intellectual disability, and is complemented by a smaller body specifically about informed consent for research (Dunn et al., 2007; Iacono, 2006; Office of the Public Advocate, 2010). I do not intend to rehearse here the complex arguments, merely to alert the reader to the controversies. A central question for ethical practice is whether the person is able to give informed consent. Do they understand what involvement will entail and the potential implications? Qualitative research that compiles rich descriptive data about people's lives, from observations or interviews creates a minefield of potentially negative consequences that are difficult to explain, predict or demonstrate as being fully understood (Pitts & Smith, 2007). For example, although de-identification strategies may disguise informants to the wider world, informants will always be identified by those who know them well or have been closely involved in a project. This raises questions to be considered during consent processes, such as:

- Will you be comfortable talking about your family and friends?
- What might be the negative implications?
- How will your siblings, nieces or care workers react when they see in print not only details of their relationship with you, but also of their relationships with each other?
- What will be the implications for your care staff or yourself if you talk about any less than ideal or even abusive care practices that might have taken place, and which would then be exposed in a research report?

I am aware of one instance that arose where the publication of interview data severely damaged the relationship between the informant and a close relative, creating a significant gap in his social network for many years.

Requirements to demonstrate capacity to consent, and who might do so on another's behalf differ between jurisdictions. The Australian research community is governed by the National Health and Medical Research Council (NHMRC) guidelines on Human Research, which stipulate that if a person cannot give consent it must be given by their guardian or person with legal authority (NHMRC, 2009, para. 4.5.5). The issue here that bedevils Australian researchers is evident to those familiar with the United Nations Convention on the Rights of Persons with Disabilities which, although subject to differing interpretations, opposes removal of decision-making powers from a person, in favour of graduated supported decision-making (Office of the Public Advocate, 2010). Only a very few people with disabilities have a guardian and a legally appointed substitute decision-maker and then only for specific types of decisions at specific points in time. This approach is reflected in the least restrictive approach adopted in Victoria, where reliance is placed on informal rather than formal decision-making processes. How does one resolve the situation where a person cannot give informed consent but there is no legally appointed person who can do so on their behalf? Some researchers use the concept of assent rather than consent, particularly for observational studies, that is, if the person does not strongly object to an observer's presence they are assumed to assent to participation. Others fudge the issue and use phrases such as 'next of kin' which, though commonly accepted, fails to acknowledge that parents do not automatically assume decision-making powers in respect of their adult children. In an ongoing Australian study with colleagues at the Tizard Centre we have used a procedure developed by them to work through consent issues. The first part asks a series of questions to support a decision about a person's capacity to consent.

- Does the person understand the information? For example, when you explain what is involved in participation, does the person show signs of understanding/lack of understanding?
- Does the person retain the information? For example, when you ask the person if they have thought about what you were discussing earlier can they tell you what it was?
- Does the person use the information to reach a decision? For instance, is the person able to give reasons for their decision that are related to the likely benefits/risks, and the alternatives?
- Does the person communicate a decision? For example, does the person tell you clearly that they do/do not wish to participate?

The second part sets out steps to reach a decision about participation of the person who is unable to give their consent directly:

- Are there any ways in which the individual's views about participation might be explored?
- In light of the above, what are the views of all concerned about the likely benefits/risks of participation?
- What is the consensus around what is in the person's best interests?
- If there is no consensus, what is the nature of the conflict?
- Record of decision. Research will proceed if there is consensus that it is in the best interests of the individual. Research will not proceed if the consensus is that it is not in the best interests of the individual, or if there is no consensus.

Approaches used with people with mild intellectual disability have been to hold briefing meetings with potentially interested participants so they can talk with each other as well as the researcher about what the research might involve (Frawley & Bigby, 2011). As Griffin and Balandin (2004,

p. 69) suggest 'consent is an interactive construct and will vary according to the complexity of the information presented, whom it is presented by, effort put into communicating it, and the setting in which the information is conveyed'. Gaining informed consent from participants will pose challenges for every project that involves older people with intellectual disability, and the solutions will differ and depend on the nature of the project. Too often however, people are excluded because it is too hard to adapt to their unique support needs, which in my view is not acceptable.

Data collection

People with mild or moderate intellectual disability, who use language to communicate are certainly able to participate in research interviews, but appropriate adaptation must occur that takes account of their degree of intellectual impairment. This group will often have difficulty remembering details, dates or sequences of events, and speaking fluently or at length about abstract concepts. People with intellectual disability often use acquiescence and 'passing', as strategies to get by and mask a lack of understanding (Edgerton, 1967; Edgerton et al., 1984; Finlay & Lyons, 2002). Such strategies can mean that if simple yes, no, closed or multiple choice questions are used, people with intellectual disability will pick the last answer, agree with the interviewer or say what they think the interviewer wants to hear. An open-ended conversational style, using an aide memoir, but following the narrative the person himself/herself wants to tell, is suggested as a good interviewing strategy (Booth & Booth, 1996, 2000). Building rapport and trust, being flexible in approach, and attention to a safe and comfortable environment are all central to good interviewing practice. This means considering who is present during interviews and where they are conducted, and acknowledging the potentially constraining influence of family or staff.

Experience suggests too, that little is gained from one-off interviews where there is no prior rapport or trust between interviewer and interviewee. Strategies that work well are illustrated by the approach taken by Grant et al. (1995), where interviewer and interviewee spent short periods of time together over a number of weeks, in social settings in the community away from spaces shared with others. Gaining familiarity with a person's past and present situation through prior interviews with others who know them well may be useful in helping construct a scaffold to ensure an interviewer provides prompts relevant to the interviewee. For example, an interviewer knowing the names of the interviewee's friends, what others see as their interests, or an outline of their life story may help reduce the abstract nature of questions about what a person might like to do if they were to retire. However, this must be done in a flexible and facilitative manner that foreshadows and prompts issues for discussion and that leaves open possibilities for the unknown to emerge. Scaffolding interviews in this way help to support people with intellectual disability deal with the abstract concepts of time, the future and past that they find difficult.

Joint interviews or group discussion among people with shared interests or experiences, can generate ideas and stimulate memories, as well as reducing the pressure individuals may experience with more intensive approaches. An approach such as this can also be useful at the preliminary stages of research and aid the development of interviewing resources via research collaborations between academics and people with an intellectual disability. For example, in a collaborative study with self-advocates, the academic partners constructed, from an analysis of literature and internal documents, a timeline of key landmarks in government policy and organisational history. This was discussed as a group with the self-advocate collaborators, who at first thought they had little to contribute but then proceeded to talk for an hour fleshing out the skeleton the academic work had provided, with their memories and stories. Over several sessions, together we pieced a pictorial 'Key Moments' document that became a tool to be used as a memory trigger for interviewing other self-advocates and allies in the project (Bigby et al., 2010a). Another strategy used commonly, particularly in the UK, has been the use of role play in the form of sketches, to both stimulate discussion in group interviews about sensitive topics, and to enable people to have an additional medium with which to explain or express feelings about things that have happened to them. The work of the Growing Old with Learning Disabilities (GOLD) group is an example of this, where a long standing group of older people have captured, using drama, issues experienced as part of ageing (Blackman & Brooks, 2008).

Interviews alone are a poor means of data collection with people with intellectual disabilities, particularly if research is focused on what happens to people in their everyday lives, rather than what they would like to happen or how they perceive reality. Ethnographic methods, participant or non-participant observations, where field notes capture detailed minutia of daily life and interactions, are important methods to supplement interviews, and often provide a more in-depth picture of a person's life than can be provided by the participant. Edgerton's classic *The Cloak of Competence* (1967) and its sequels, which follow the lives of a group of people who lived in an institution remain the seminal ethnographic work in this field.

The final installment *'I've Seen It All' Lives of Older Persons with Mental Retardation in the Community* (Edgerton & Gaston, 1991) vividly illustrates the life of a small group of older men, and the costs they bear in terms of health and loneliness in retaining their independence and autonomy in everyday life. For many people with intellectual disabilities with very limited communication skills, or more severe impairment, this type of ethnographic study is the best way to capture a detailed perspective on lives unencumbered by notions of best interests or protection. Increasingly however, as people live in communal settings and move about the community, ethnographers must grapple with ethical issues such as who is included in the field of observations, and from whom informed consent must be sought.

Multiple informants

A less intrusive and resource intensive strategy to gain insights into the lives of older people with intellectual disability and family or service system responses, is to use single or multiple proxy informants. Recruiting a cluster of informants around each older person is a common strategy. For example, conducting interviews with a family member, a front-line worker and a more senior manager from each of their services as well as where feasible, the person with intellectual disability too, is a commonly used strategy. A snowball strategy can also be used to identify and recruit members of a person's informal network. The purpose is not triangulation as would be the case in more positivistic research, but involvement of people who cannot represent their own views, the gaining of multiple perspectives on the life of people with intellectual disability and issues associated with ageing, as well as factual or chronological information that can be used to scaffold other data collection strategies.

Using multiple informants can highlight very differing ideas that may be held about ageing, about service quality, the 'best interests' of an older person with intellectual disability, or strategies employed to control people's lives. For example, in a recent study, comments by older siblings illustrated the strategies staff used to avoid conflict about what might be in the best interests of their clients (Bigby et al., 2010b). A sibling said:

> I was never actually informed about this until it was too late....I mean I was never actually informed of this until very late in the piece and so I hadn't been consulted or advised of anything, just came like a

bolt out of the blue, oh Charlie 's going to America (Field note from Webber et al., 2010).

In the same study interviews with staff confirmed the widespread use of this type of underhand strategy to retain control of decisions about resident's lives. For example, one staff member said: 'he [resident with intellectual disability] has been fitted with a hearing aid... the house manager hadn't told her [sister] and said she would do so when they were fitted and he was used to them' (ibid.).

By drawing on the views of siblings and other informal network members separately from those of elderly parents, various studies have shed light on the complex dynamics at play in families where there is an ageing person with intellectual disability. For example, the difficulties siblings have in challenging perceived overprotection of their sibling with intellectual disability by an elderly parent. One man said about his mother's relationship with his brother with intellectual disability:

> Phillip's her little boy...mother was too dominant and did too much for him. She's done a marvelous job, looked after him very well. But only when you get him away from her do you realise she has made it extremely difficult for anyone else. You can't criticize her for the care she gave him. He would have had a much happier life with her than going into one of those homes (Bigby, 2000, p. 116).

Other siblings talked about the strategies they used to avoid confronting parents or forcing them to have to come to terms with relinquishing care of their adult with intellectual disability. 'Unbeknownst to my mother I went round and had a look at nursing homes for him [brother]. I didn't want to be sneaky but I didn't want to upset her. The best thing was quietly to have a look' (Bigby, 2000, p. 86). Use of multiple informants has provided significant insights into the often unacknowledged caring tasks and restrictions that result for adults with an intellectual disability, from continued living with an elderly parent. George's brother said about his mother, 'she was as dependant on George as he was on her. It was a knife edge situation'. George himself said:

> I couldn't have been able to come to functions here or at the church. If I went to the cricket all day Saturday I'd come home and find mum on the floor and she wouldn't be able to get any help, she wouldn't be able to ring up. I just couldn't get away and leave her on her own.

> I'd have to stay with my mum. I couldn't do the things which I would like to do (Bigby, 2000, p. 76).

One potential drawback of relying on multiple or proxy informants is the inherent sampling bias this can create towards research participants who are service users or have involved family members. As one sibling remarked in a recent study of older people with intellectual disability living in group homes: 'Makes you wonder how people get on that don't have family' (Bigby et al., 2012). Indeed it does, and for some of the reasons I have already discussed we know very little about this.

This type of research points to the value of using multiple rather than a single informant to understand ageing families, and can provide valuable insights for practitioners who work with them. Increasingly, compiling life stories is a suggested strategy in work with older families in order to ensure the person's history is not lost when their parents die. A fairly grim picture on later life for people with intellectual disability is often painted by relying on data about the anxieties parents have about the future. But research on later life that helps people to reflect on their life since the death of their parents is important too, in shedding a more critical light on the frequently proposed rational ideas about parents planning for the future. For example, in an early paper from my 2000 study I reported examples of elderly people with intellectual disability whose horizons had broadened after the death of their parent in ways that could never have been envisaged. The case of Rod and Isobel is just one poignant example:

> Rod and Isobel married when they were in their 50s, several years after their respective parents had died. They lived together independently in the community for about 24 years, supported informally by members of their church community. The family friend who managed Rod's affairs since his parents died said, 'of course they [his parents] never expected Rod to marry. That's the last thing in the world they ever imagined' (Bigby, 1997, p. 100).

In a similar vein people with intellectual disabilities often have a clear sense of what they want to happen in the future, which may not be considered when plans about their future plans are made. For example, Malcolm (a service user), who lived in a rural accommodation service said: 'I'd like to live with my girlfriend that's what I've got in the back of my mind. I'd like to have my own house with her. I've worked out how I can move…but it's not happening.' Rachel, a service user who lived with her elderly parents, had a vision of what she wanted for her future: 'I'd like to move somewhere where there's no stairs and somewhere where there [are] lots of shops and trains...I've had too much of here...Forty years nearly forty-one' (Bigby & Knox, 2009, p. 224).

These examples point to the fundamental need to involve people with intellectual disability in research in order to uncover their hopes for later life, as well as those of others, and to involve them in preparing for their own futures. Moving on from the micro level of ensuring the voices of people with intellectual disability are heard in research about their lives, the last part of this chapter widens its focus to consider the value of a qualitative approach in understanding programme design or service systems, and including people as partners in research rather than simply informants.

Applied research

Much applied social research is about things that make a difference to people's lives: family support, attitudes, policy formation and its implementation through programmes and service delivery. Action research and participatory approaches are interested in processes of change, how and why particular approaches work, as well as, involving and empowering service users or staff. The aim of traditional intervention studies, such as those revered by the Cochrane Collaboration is to isolate variables and demonstrate the impact of particular interventions. The real world of human services however, is complex and messy, with multiple contextual variables as well as formal intervention affecting outcomes. Ethically, few situations lend themselves to controlled experimental designs. Within this world however, qualitative research has a significant role to play in uncovering the complexities involved in effective programmes and methods of intervention. This type of research requires researchers to be skilled in multiple roles. For example, in a recent study on planning for the retirement of employees in supported employment, the purpose was to evaluate the outcomes for older people that flowed from their inclusion in a chosen community-based volunteering or social group, where another member had been trained to mentor them. However to get to the stage where the role of mentors and social inclusion could be observed, retirement planning with individuals, and location of community groups which offered activities that matched their interests, was necessary. The study design overestimated the capacity of the supported employment service from which people were retiring to undertake these tasks, and the multiple steps and negotiations involved. By one of the researchers assuming the role of *de facto* case manager, the study demonstrated the nature of this role as an essential but missing element in the design of retirement programmes for older people with intellectual disability (Wilson et al., 2010).

One feature of researching the lives of older people with intellectual disability is the disconnection often found between community and staff attitudes and those portrayed in overarching policies such as the Convention on the Rights of Persons with Disabilities. This is often reflected in an implicit hierarchy of people with disabilities, with those with intellectual disability at the bottom, which in turn means that concepts such as community inclusion, participation and access are deemed to be inapplicable, or interpreted and applied in a manner that denies consideration of issues relevant to those with intellectual disability (Bigby et al., 2009). This can mean that action researchers may have to assume an advocacy role with doubtful staff to promote the right of people with intellectual disability to be included in the community. In an action research study of community group homes for instance, after more than 18 months of working with staff to support social inclusion of residents, our finding suggested that despite strong advocacy from the researcher, some staff were still uncertain about the feasibility and validity of this idea (Clement & Bigby, 2009).

The adage from the disability movement 'nothing about us without us' has spawned many strategies for including people with intellectual disability in research, not only as informants but also in the processes of problem formulation and conduct (Walmsley, 2010). Skills and approaches to 'inclusive research' are evolving. Principles based on theory are being questioned, as researchers attempt to apply them in the field (Bigby & Frawley, 2010), and calls are made to lift the veil on what occurs under the guise of inclusive practices. Our recent work on the history of self-advocacy has drawn attention to the immense value to rigor and quality of research that flows from collaboration between academics and people with intellectual disability. However, rather than teaching people with intellectual disability to imitate the skills of academic researchers, or just thinking in terms of support or easy to read materials, we have drawn on Seale and Nind's (2010) work on the concept of access. They suggest that access is a process rather than an absolute concept. It involves much more than just getting through the door, and must have a quality dimension which extends into the realm of what happens thereafter. Thus, for a person with intellectual disability, access to research meetings for example is 'to be able to know' – to both understand the situation and to gain some benefit from being there and participating. Therefore we paid attention to scaffolding access for self-advocates by adapting research processes. For example, we created a cognitively accessible space in which the group operated, where self-advocates possessed an accepted and valued place, and did not feel excluded by complex language or concepts. We invited interviewees into this space and conducted group-based interviews in which self-advocates participated on their own terms, based on their own knowledge and skills, and on information generated through joint work in earlier stages of the research, such as the Key Moments documents (see above). Unlike other inclusive research groups, we have acknowledged the legitimacy of using non-accessible space, tasks and documents to undertake the preparatory work necessary to scaffolding access for self-advocates. As I outlined earlier in the paper, the 'Key Moments' book of photos used in interviews was not just a random selection. It originated in an academic analysis of key policies and events in the history of the organisation presented to the group which, with the knowledge of members, was amended and converted into a useful prompt for interviewers and interviewees alike. We drew on Mauthner and Doucet's (2008) ideas about research collaboration, which recognises the diverse skills and contributions made by members of research groups, as well as the shared outcomes and purposes sought. In our case, all parties sought to further the cause of self-advocacy, the academics needed to write journal papers, and the self-advocates wanted to preserve their story and pass it on for the next generation, rather than to other academics. Unlike some research groups however, we endeavoured to value all contributions equally, and give as much attention to outcomes sought by self-advocates as to those of academics (Bigby et al., 2010a). Like most research with people with intellectual disability, research collaboration requires significant time and more than usual resources, which is rarely recognised by research funding bodies or universities. Just as with any colleagues, collaborative research with self-advocates relies on relationships of trust, developed over time, and based on shared experiences, regular contact and reliability.

This chapter has given a brief overview of the disadvantageous position of people ageing with intellectual disability and set out some of the particular research challenges in undertaking qualitative research with this group. Unlike many older people, this group will find it difficult to be informants about their own lives. I hope the chapter has provided some helpful ideas, and demonstrated the value of persevering to ensure the perspectives of people with intellectual disability about their own ageing, as well as the sometimes conflicting views of others, are captured. Such perspectives are important to support advocacy from the standpoint of people with intellectual disabilities, and to ensure appropriate support is provided that will enable this group of disadvantaged and marginalised older people to age well.

Annotated reading list

Edgerton, R. & Gaston, M. (1991) *'I've Seen It All': Lives of Older Persons with Mental Retardation in the Community* (Baltimore, Maryland: Paul H Brookes).

This text is a follow up to Edgerton's 1967 classic *Cloak of Competence* which was an ethnographic study of a group of people with intellectual disability released from a US institution in the 1960s. It details the later life experiences of this group.

Bigby, C. (2004) *Aging with a Lifelong Disability: Policy, Program and Practice Issues for Professionals* (London: Jessica Kingsley).

This text provides a comprehensive overview of the biopsychosocial issues confronting people with intellectual disability as they age.

Blackman, N. & Brooks, M. (2008) *Dementia and People with Learning Disabilities Valuing Relationships* (London: Respond in partnership with the Gold Group).

This book and the two CDs that accompany it present the perspectives of people with intellectual disability about ageing, and demonstrate the power of using drama to convey experiences.

Manning, C. (2008) *Bye-Bye Charlie: Stories from the Vanishing World of Kew Cottages* (Sydney: University of New South Wales Press).

This text uses oral history with staff and older residents to portray the history of the first institutions built in Australia specifically for people with intellectual disability.

Walmsley, J. (2010) 'Research and emancipation: Prospects and problems' in G. Grant, M. Richardson & J. Murphy (eds) *Learning Disability: A Lifecycle Approach to Valuing People*, pp. 489–502, 2nd edn (Maidenhead: Open University Press).

This chapter provides an overview of debates about including people with intellectual disability as partners in research.

Walmsley, J. (1998) 'Life history interviews with people with learning disabilities' in R. Perks & A. Thompson (eds) *The Oral History Reader* (London: Routledge).

This chapter provides an introduction to undertaking life history work with people with intellectual disability, and refers to key texts in this field.

References

Bigby, C. (1994) 'A demographic analysis of older people with intellectual disability registered with community services Victoria', *Australia and New Zealand Journal of Developmental Disabilities*, 19, 1–10.

Bigby, C. (1995) 'Is there a hidden population of older people with intellectual disability and from whom are they hidden? Lessons from a recent case-finding study', *Australia and New Zealand Journal of Developmental Disabilities*, 20, 15–24.

Bigby, C. (1997) 'Later life for adults with intellectual disability: A time of opportunity and vulnerability', *Journal of Intellectual and Developmental Disability*, 22, 97–108.

Bigby, C. (2000) *Moving On Without Parents: Planning, Transitions and Sources of Support for Older Adults with Intellectual Disabilities* (New South Wales/Baltimore: Mclennan+Petty/P H Brookes).

Bigby, C. (2004) *Aging with a Lifelong Disability: Policy, Program and Practice Issues for Professionals* (London: Jessica Kingsley).

Bigby, C., Clement, T., Mansell, J. & Beadle-Brown, J. (2009). '"It's pretty hard with our ones, they can't talk, the more able bodied can participate": Staff attitudes about the applicability of disability policies to people with severe and profound intellectual disabilities', *Journal of Intellectual Disability Research*, 54 (4) 363–76.

Bigby, C. & Frawley, P. (2010) 'Reflections on doing inclusive research in the "Making life good in the community" study in Australia', *Journal of Intellectual and Developmental Disability*, 35 (2) 53–61.

Bigby, C. & Knox, M. (2009) '"I want to see the Queen", The service experiences of older adults with intellectual disability', *Australian Social Work*, 62 (2) 216–31.

Bigby, C., Ramcharan, P. & Frawley, P. (2010a) 'Researching self advocacy: The first 3 years of an inclusive study by self advocates and academics', *Journal of Applied Research on Intellectual Disability*, 23 (5) 453.

Bigby, C., Webber, R. & Bowers, B. (2010b) 'Family roles and relationships of older adults with ID resident in group homes', *Journal of Applied Research on Intellectual Disability*, 23 (5) 417.

Bigby, C., Webber, R. & Bowers, B. (2012). '"We all kind of work in together": Relationships in later life between group home residents with intellectual disability, their siblings and staff' in C. Bigby & C. Fyffe *Family and Services Working Together. Proceedings of the 2011 Roundtable on Intellectual Disability Policy* (Department of Social Work, LaTrobe University).

Bigby, C., Webber, R., McKenzie-Green, B. & Bowers, B. (2008) 'A survey of people with intellectual disabilities living in residential aged care facilities in Victoria', *Journal of Intellectual Disability Research*, 52, 404–14.

Blackman, N. & Brooks, M. (2008) *Dementia and People with Learning Disabilities Valuing Relationships* (London: Respond in partnership with the Gold Group).

Booth, T. & Booth, W. (1996) 'Sounds of silence: Narrative research with inarticulate subjects', *Disability and Society*, 11 (1) 55–70.

Booth, T. & Booth, W. (2000) 'Against the odds: Growing up with parents who have learning difficulties', *Mental Retardation*, 38 (1) 1–14.

Clement, T. & Bigby, C. (2009) 'Breaking out of a distinct social space: Reflections on supporting community participation for people with severe and profound intellectual disability', *Journal of Applied Research in Intellectual Disabilities*, 22 (3) 264–75.

Dunn, M.C., Clare, I.C.H., Holland, A.J. & Gunn, M.J. (2007) 'Constructing and reconstructing "best interests": An interpretative examination of substitute decision-making under the Mental Capacity Act 2005', *Journal of Social Welfare and Family Law*, 29 (2) 117–33.

Edgerton, R. (1967) *The Cloak of Competence* (Los Angeles CA: University of California Press).

Edgerton, R., Bollonger, M. & Herr, B. (1984) 'The cloak of competence: After two decades', *American Journal on Mental Deficiency*, 88, 345–51.

Edgerton, R. & Gaston, M. (1991) *'I've Seen It All' Lives of Older Persons with Mental Retardation in the Community* (Baltimore, Maryland: Paul H Brookes).

Emerson, E., Hatton, C., Felce, D. & Murphy, G. (2001) *Learning Disabilities: The Fundamental Facts* (London: The Foundation for People with Learning Disabilities).

Emerson, E., Malam, S., Davies, I. & Spencer, K. (2005) *Adults with Learning Difficulties in England 2003/4* (London: Department of Health).

Finlay, W. & Lyons, E. (2002) 'Acquiescence in interviews with people who have mental retardation', *Mental Retardation,* 40, 14–29.

Frawley, P. & Bigby, C. (2011) 'Inclusion in political and public life: The experiences of people with intellectual disability on government disability advisory bodies in Australia', *Journal of Intellectual and Developmental Disability*, 36 (1) 27–38.

Grant, C., McGrath, M. & Ramcharan, P. (1995) 'Community inclusion of older adults with learning disabilities', *Care in Place – International Journal of Network and Community*, 2 (1) 29–44.

Griffin, T. & Balandin, S. (2004) 'Ethical research involving people with intellectual disabilities' in E. Emerson, C. Hatton & T. Parmenter (eds) *International Handbook of Applied Research in Intellectual Disabilities*, pp. 61–82 (Chicester: John Wiley and Sons).

Haveman, M., Heller, T., Lee, L., Maaskant, M., Shooshtari, S. & Styrdom, A. (2010) 'Major health risks in aging persons with intellectual disabilities: An overview of recent studies', *Journal of Policy and Practice in Intellectual Disabilities*, 7 (1) 59–69.

Hogg, J., Moss, S. & Cooke, D. (1988) *Ageing and Mental Handicap* (London: Croom Helm).

Iacono, T. (2006) 'Ethical challenges and complexities of including people with intellectual disability as participants in research', *Journal of Intellectual and Developmental Disability*, 31 (3) 173–9.

Jacobson, J., Sutton, M. & Janicki, M. (1985) 'Demographic and characteristics of aging and aged mentally retarded persons' in M. Janicki & H. Wisniewski (eds) *Aging and Developmental Disabilities: Issues and Approaches*, pp. 115–43 (Baltimore: Brookes).

Janicki, M., McCallion, P., Force, L., Bishop, K. & LePore, P. (1998) 'Area agency on aging and assistance for households with older carers of adults with a developmental disability', *Journal of Aging and Social Policy,* 10 (1) 13–36.

Krauss, M. & Seltzer, M. (1986) 'Comparison of elderly and adult mentally retarded persons in community and institutional settings', *American Journal of Mental Deficiency*, 91, 237–43.

Llewellyn, G., McConnell, D., Gething, L., Cant, R. & Kendig, H. (2010) 'Health status and coping strategies among older parent-carers of adults with intellectual disabilities in an Australian sample', *Research in Developmental Disabilities*, 31 (6) 1176–86.

Marmot, M. & Wilkinson, R. (1999) *Social Determinants of Health* (Oxford: Oxford University Press).

Mauthner, N. & Doucet, A. (2008) '"Knowledge once divided can be hard to put together again": An epistemological critique of collaborative and team-based research practices', *Sociology*, 42 (5) 971–85.

NHMRC (2009) National Health and Medical Research Council Annual Report 2009–10, Australian Government (available online: www.nhmrc.gov.au/guidelines/publications/nh144).

Office of the Public Advocate, South Australia (2010) *Annual Report 2010* (Adelaide: Government of South Australia).

Pitts, M. & Smith, A. (2007) *Researching the Margins: Strategies for Ethical and Rigorous Research with Marginalized Communities* (New York: Palgrave Macmillan).

Reinders, H.S. (2002) '"The good life for citizens" intellectual disability', *Journal of Intellectual Disability Research*, 46 (1) 1–5.

Seale, J. & Nind, M. (eds) (2010) *Understanding and Promoting Access for People with Learning Difficulties: Seeing the Opportunities and Challenges of Risk* (London: Routledge).

Thompson, D.J., Ryrie, I. & Wright, S. (2004) 'People with intellectual disabilities living in generic residential services for older people in the UK', *Journal of Applied Research in Intellectual Disabilities*, 17, 101–8.

Torr, J., Strydom, A., Patti, P. & Jokinen, N. (2010) 'Ageing in Down syndrome: Morbidity and mortality', *Journal of Policy and Practice in Intellectual Disabilities*, 7 (1) 70–81.

United Nations (2006) *Convention of the Rights of Persons with Disabilities and Optional Protocol*, http://www.un.org/disabilities/documents/convention/convoptprot-e.pdf, date accessed 14 October 2008.

Walmsley, J. (2010) 'Research and emancipation: Prospects and problems' in G. Grant, M. Richardson & J. Murphy (eds) *Learning Disability: A Lifecycle Approach to Valuing People*, 2nd edn, pp. 489–502 (Maidenhead: Open University Press).

Webber, R., Bigby, C. & Bowers, B. (2010) 'Increasing organisational capacity of community residential units to facilitate ageing in place for people with intellectual disability', ARC Linkage Grant (unpublished data).

Wilson, N.J., Balandin, S., Stancliffe, R.J., Bigby, C. & Craig, D. (2010) 'The applied researcher as transition to retirement case manager', paper presented at the NDS Employment Forum: Melbourne, September.

WHO (World Health Organisation) (2002) *Active Ageing a Policy Framework* http://whqlibdoc.who.int/hq/2002/WHO_NMH_NPH_02.8.pdf, date accessed 18 December 2010.

6
Interviewing Older Men
Miranda Leontowitsch

Introduction

Qualitative research into the experiences of older men is still a rarity. More generally, studies on ageing and gender have mostly addressed the experiences of women and gay men, while studies of masculinities have tended to focus on younger men, leaving issues of masculinity and gender in older, heterosexual men relatively under-researched. Paradoxically, ageing men are far from being on the margins of contemporary society. Film, magazines and literary fiction do not shy away from portraying the lives and experiences of older men and news broadcasts, television shows, weather forecasts and documentaries provide us with an abundance of senior male voices and figures. As Hearn (1995) points out, these representations focus on successful men and portray them as the norm. This in turn has contributed to seeing older men as unproblematic and not in need of much attention. The small literature available on men in later life suggests that ageing provides men with both challenges and opportunities for their identity, wellbeing and relationships (Suen, 2010, p. 199). Qualitative interviewing is particularly apt for exploring under-researched groups as well as sensitive topics. Despite an emerging interest in the experiences of ageing men, little work reflects how social researchers interview this group of men. This chapter will look at the reasons as to why heterosexual ageing men remain largely at the margins of research, highlight issues arising from qualitative studies into researching their experiences, and provide a reflection on my own research experiences of conducting interviews with men over 50. This reflection, in combination with the literature covered, will examine the dynamics of interview settings where a mismatch exists in age and sex between the interviewer and the interviewee.

Research on ageing men

Edward Thompson (1994, 2006) identified several reasons for why ageing men are absent from research, namely, the smaller number of older men in comparison to older women, gerontologist resistance to distinguishing between sex and gender, the political economy perspective, compassionate ageism, homogenised later life, and ageing as a negation of masculinity. Nearly 20 years have passed since Thompson's publication, and we continue to see an increase in life expectancy across many countries, with the age gap closing between men and women. In the United Kingdom (UK) women continue to live longer than men, but in the past 27 years the gap has narrowed from 6 to 4.2 years (ONS, 2010). In general the political economy perspective focused on poverty and ageing as a residual category, and in doing so concentrated on old women who, due to no or short working lives possessed little in terms of pensions, contributed a significant amount of care for husbands and other family members, outlived husbands, and as a consequence lived in poverty. For these reasons ageing was often considered to be fundamentally an experience of women (Arber & Ginn, 1991). In terms of research, there is a sociological tradition of researching the disadvantaged rather than the advantaged and this rings particularly true for qualitative research (Hammersley, 2008). This has meant that the vast majority of research has focused on women, and in particular on the hardship of women. Men have not been considered 'worthy' of investigation as they have not been seen to be a marginalised group in need of emancipation. The relative comfort of older men's lives (e.g. paid mortgage, living with spouse, cared for rather than caring) underlines this argument (Calasanti, 2004). Although the political economy focus has provided valuable insights into the plight of older women and initiatives for social policy, it has also led to viewing older people as a homogenous group for whom compassion needs to be shown. The political economy approach has been slow to recognise the significant changes to later life witnessed over the past 40 years (Gilleard & Higgs, 2000), which have introduced an increase in the number of occupational pensions, and the prevalence of house ownership, women's employment, improvements in health and health care, and the consumer culture engrossing all life phases (see Chapter 1). This is not to suggest that poverty, social exclusion and inequalities in later life has been eradicated, but it does call for research that takes these changed parameters into account and acknowledges the heterogeneity of older people.

Older men and masculinities

In contrast to the absence of older heterosexual men in social research, an abundance of qualitative research has been conducted on ageing gay men, such as Lee's work on the sexual and ageing identities of gay men (2008). Sociological work has recognised the contribution gay liberation and gay scholarship has made to the critical analysis of previously accepted notions of sexuality and gender as being 'natural' (Hearn, 1990, 1995). Together with modern feminism this has led to the 'recognition that the dominant forms of men and masculinities are themselves not merely "natural" and unchangeable. (...) Thus men and masculinities are not seen as unproblematic, but as social constructions which need to be explored [and] analysed' (1990, p. ix). This position, in combination with the recognition of significant changes to later life has encouraged research on older men, and provided a lens through which to identify hidden complexities. In the past 12 years qualitative research into the lives of older men has looked at: the experience of widowers and how this reinforces men's sense of masculinity through the bereavement process (Bennett, 2007; van den Hoonaard, 2010), masculinity and caregiving (Robeiro et al., 2007), being a grandfather (Mann, 2007; Reich, 2007), the construction of a positive ageing identity (Phoenix & Sparkes, 2009), their health (Davidson & Meadows, 2010), how men perceive their ageing bodies (Hurd Clark et al., 2008), the experience of growing old in terms of home and leisure activities (Russell and Porter, 2003; Russell, 2007), issues around independence (Smith et al., 2007; Stephenson et al., 1999), being single (Suen, 2011), being a runner (Tulle, 2008) and sexuality (Gott & Hinchliff, 2003; Potts et al., 2006). Not all these studies are purely about men's experiences as several look at women and men. But what is central to all is the role of masculinity, and the gendered relationships between men and women that persist throughout the lifecourse.

Thus some progress has been made since Thompson's critique in 1994, that social gerontology does not distinguished between sex and gender, and thus ignores older men's masculinities. Or that ageing is seen as a negation of masculinity, where older men are portrayed as obsolete in a culture cherishing power. Moreover Thompson (1994) accuses the media of playing a large role in presenting men as part of a 'grandparent' generation, sedentary, resting and asexual. Hearn (1995) takes a more nuanced approach and analyses how older men are portrayed through a number of genres in the press and mass media, which concentrate on previously successful men in positions of power and authority. It is for this reason that we see and hear more of older men than women. These two perspectives go some way in showing how older men can be seen in different ways and how this has changed over time. With the emergence of the 'third age' (Laslett, 1989; Gilleard & Higgs, 2005) as a phase of life past employment and child rearing, in which self-fulfilment is achieved through travel, education and finding new social roles, the public portrayal of later life has changed further. The new active older person is associated with both men and women. Indeed this has been promoted and exaggerated to such an extent that it now problematises those older people who do not appear to be embracing an active lifestyle (Asquith, 2009; Katz & Marshall, 2003). Framed this way, images of age-specific and cohort-specific masculinity are rising to a threshold of public consciousness. However, changes in public perceptions of later life are never absolute but rather seamless and gradual. With a continuously growing ageing population which is also becoming older, different cohorts of older and younger people will associate ageing men with different ideas in terms of gendered roles, masculinities and images. Hearn and Pringle (2006) make a timely observation when discussing the creation of more gendered images of men, and the openness to debate of previously taken-for-granted male powers and authority: 'The paradox is that men and masculinities are now more talked about than ever before when it is much less clear what and how they are or should become' (p. 3). The research on older men so far suggests that this is true for both young and old men. Thus, qualitative research on ageing men across later life needs to further investigate notions of perceived and acted masculinities. Moreover, if these are central to men growing older it is also central to how they construct their identity and present themselves to others, and in particular, to researchers. It is important to understand how different ideas of gender and masculinity are intrinsic to the research process, and how they shape the research relationship between interviewer and interviewee and the data collected (Johnson, 2001).

Some methodological issues

Current qualitative research provides strong evidence that older people themselves construct ageing as a gendered phenomenon. 'Whether the topic under investigation is emotional attachment to the home, leisure activities in retirement or socio-emotional engagement in later life, qualitative research reveals differences in the meanings that older men and women confer on it' (Russell, 2007, p. 187). These differences are shaped by the different gendered roles men and women have. With more social researchers recognising the centrality of gender to understanding how older people experience and make sense of later life, critical views on how the gender, age and sex of both the interviewer and interviewee affect the interview are being voiced. In addition, theoretical sampling and the need for reflexive research have been identified as methodological issues. I will consider all of these in turn.

Theoretical sampling

The 'social problem' focus, with its emphasis on poverty and lack of agency, in combination with theoretical sampling is one such methodological concern (Russell, 2007), and can over-represent the negative face of ageing. In theoretical sampling the interviewees are purposefully selected on the basis of their 'authority' on a certain issue. In conducting such research, potential participants are selected from settings that yield rich ethnographic data, such as nursing homes or other care services. These settings are predominantly frequented by women, but findings are often glossed as the views of 'residents' or 'clients' (Russell, 2007, p. 187). Russell identifies three qualitative studies looking at institutional quality of life where the voices of older residents could not be identified by sex. Even where the political economy perspective does not provide the backdrop to research, qualitative research's general reliance on theoretical sampling can lead to samples with more women than men. In addition, a reluctance of older men to participate socially makes it more difficult to target older men for sampling purposes (Suen, 2010). My own experience of recruiting participants aged 60+ from health food shops and pharmacies for a study on self-care using non-prescription and complementary medicines (Leontowitsch et al., 2010) resulted in a far larger group of female participants. It was indeed harder to persuade men to be interviewed and there were fewer men to approach. Shopping, as a still largely female gender role, may account for the latter.

Interviewer effect

Russell (2007), who has researched the differences of men and women in the lived experience of ageing, argues that there is reason to suspect that even when men are included in qualitative research, the data collected may have been impoverished. This is based on the idea that the sex of the interviewer and that of the person interviewed affects what is being said in the interview. This idea was put to test by Stephenson et al. (1999), who looked at the effects of different interviewers on the information provided by older men and women when interviewed about home-making services and service delivery. The researchers divided their sample of 39 participants (19 men and 20 women) into two sets. In the first set nine men were interviewed by a male researcher and ten women were interviewed by a female researcher. In the second set ten men were interviewed by a female researcher and ten women were interviewed by a male researcher. The same topic guide was used in all interviews. What the authors found was that when interviewed by a man, older men focused strongly on career as a key part of their identity, whereas in interviews conducted by a woman, older men emphasised their family as an important part of their identity. A similar reversed pattern could be seen in the interviews with women, where women emphasised their family life with the female interviewer, and career/education with the male interviewer. Stephenson et al. (1999) conclude that findings in the literature on independence, ageing, health and social care needs may stem from systematic, unacknowledged, interviewer-gender bias. The authors raise pertinent questions about using one-off structured interviews to assess older people's care needs, but say little about the implicit advantage of using mismatched sets of interviewer and interviewee in terms of sex to elicit rich data. I will return to the possible advantages in my reflection on interviewing older men.

Matching the age of the interviewer and the interviewee is seen as a way of securing rich data. Lee (2008), on interviewing gay older men, argues that an age match permits the understanding of key events of the 20th century and helps build rapport, because there is no need to interrupt interviewees to ask questions about historical events. An age-match of interviewee and interviewer can be achieved when the researcher is of a similar age as his/her interviewees, or, if that is not the case, when an older person is trained in research methods to conduct the interviews, an approach increasingly used in social gerontology in the UK. In an overview of these developments Biggs (2005) writes, 'It is argued that the use of older researchers to research older people may go some way to reduce biases associated with the unequal distribution of power between researchers sponsored by state organizations and those who use their services' (p. 124). In line with this, Silver (2003) points out that most of the research has been done by the non-aged, whereas 'the world of the older individual needs to be studied from the inside'. However, Biggs (2005) raises questions about whether older adults have a privileged position in terms of accessing later life experiences, by virtue of being old themselves. He draws attention to the fact that, as Silverman (1993) points out, interviews are a medium through which individuals perform a public version of their self. Moreover, there is too much respect across much qualitative research (and in line with grounded theory), for what participants say, without acknowledging the context in which statements are made. In the case of age-matched interviews about later life, this could lead to significant bias in the interview process because the interviewee may not elaborate on points they feel are commonly understood. At the same time, an older interviewer may fail to check or further explore concepts and ideas that may seem apparent. Biggs (2005) concludes that every researcher, whether young or old, needs to critically challenge his or her own generational states of mind, in order to avoid seeing what he or she expects:

> An uncritical acceptance of the views of older people simply because they are old may indicate an abdication of responsibility by the researcher or policy maker, indicating disrespect for those views, if they are not subjected to the same interrogation that would be reserved for any other data gathering exercise. (...) And although embracing same-age research holds considerable democratic potential, it may not fully solve the problems of authenticity (p. S125).

This seems to suggest that a mismatch in age between interview and interviewee has its place. I will return to the issue of different-age interview settings when reflecting on my own interview experiences.

Reflexivity

All of the above, sampling and interviewer-interviewee match or mismatch in terms of sex or age, calls for reflexivity on part of the researcher. Reflexivity is an integral part of using qualitative methods, as it enables the researcher to view his or her impact on the research setting and data collected. In so doing reflexive research takes an important role in the development of qualitative methodologies, regardless of whether these methodologies err to the naturalist or constructionist end of the ontological spectrum. However, the growing influence of postmodernist, constructionist and ethnomethodological sensibilities has intensified awareness that meaning is not a free-floating phenomenon, waiting to be captured, but is socially constituted (Holstein & Gubrium, 2003):

> Treating interviewing as a social encounter in which knowledge is constructed means that the interview is more than a simple information-

gathering operation, it's a site of, and occasion for, producing knowledge itself (ibid., p. 4).

Thus, meaning is actively and socially assembled in the interview encounter by both the interviewer and the interviewee. This understanding has been central to my research experience and conduct. It has helped me make sense of what I have collected, and shape the way I conduct interviews. The following section provides some insight into this reflective process and into what I have been able to conclude in terms of the methodological issues raised so far.

Reflecting on my own experience of interviewing older men

Over the past 11 years I have conducted qualitative interviews for a number of very different sociological research projects. In more recent years two of these projects were concerned with matters of health, which included a larger group of older participants. In addition, I was involved in a study on the experiences of early retirement among a group of men and women from higher executive positions. It is on these three studies that I will base my reflection.

Having developed my research career on a linear path since entering higher education I found myself to be at the younger end of the researcher age-spectrum, and indeed, part of the ever-growing group of young female researchers. At the time of conducting the interviews I was in my late twenties, thus anything between 25 and 42 years younger than my male participants. This, and the fact that I was a female researcher led to an increased awareness about how I was conducting interviews with a growing number of older men. Before delving into the experience of interviewing older men, I will provide a brief description of the studies (see Table 6.1), how participants were sampled, and how the studies were conducted.

'Concordance in pharmacy practice' was not focused on later life. However, health problems are more likely to occur with increased age and pharmacies are venues which dispense medicine and health advice and therefore tend to attract an older customer base. For these reasons 11 of the participating men and women were aged 54 to 81. Participants were recruited by the pharmacists of the two pharmacies taking part in the study. The study inclusion criteria were that the consultation focused on advice about over-the-counter medicines (OTCs) and was held in English. Once the pharmacist had established that the customer required a consultation for OTCs, he/she informed the customer about the study and

Table 6.1 Overview of studies involving men aged 52+ interviewed by myself

Study name	Synopsis	Total sample	Interviews with older men
Concordance in pharmacy practice	Investigating the use of concordance in consultations between pharmacists and costumers about over-the-counter medicines. Conducted in an inner-city (more deprived area) pharmacy and one in a residential (more affluent) area of London (Leontowitsch et al., 2005).	26 audio-recorded consultations with subsequent interviews (11 men, 15 women)	Five men aged 58–81
Self-care in later life	How older people use non-prescription and complementary medicines to maintain their health. Participants were sampled in independent pharmacies and health food shops in London (in both deprived and affluent areas). (Leontowitsch et al., 2010).	22 interviews (three men, 19 women)	Three men aged 65, 76, 79
Experience of early retirement	Exploring reasons for and the experience of early retirement for a group of professionals from higher executive positions. Sampled from across England (Jones et al., 2010).	20 interviews (18 men, two women)	15* men aged 52–66

*Three interviews were conducted by another member of the team.

asked whether they would mind if their consultation was recorded. After the consultation I asked each participant whether they would agree to be interviewed a few days later. Although the five men interviewed aged 58–81 lived in a more affluent residential area, their circumstances can be described as modest.

For the self-care study I approached customers who looked 60+ in five pharmacies and four health food shops that had agreed to be part of the study. The study inclusion criteria were that participants were aged 60 or over, had purchased or were interested in non-prescription or complementary medicine and spoke English. As it was difficult to establish someone's age by looking at them I also approached people who looked younger than 60 to ensure the sample was not biased to the older age spectrum (who were more easily identifiable as 60+). Two of the men interviewed lived on a state pension, whereas the third was more affluent.

Participants for the early retirement study were sampled through The Retired Executive Action Clearing House (REACH), a voluntary organisation that tries to match charities with retirees from professional backgrounds, by snowball sampling and through the Life Academy that offers courses on planning for retirement. Potential participants needed to have retired from a senior managerial position between the ages of 53 and 57.

My reflection identified four themes that emerged from conducting these interviews and that say something about conducting interviews as a female researcher who is younger than her male participants. They are: (a) gentleman-like behaviour, (b) generativity, (c) fatherly interest, and (d) the role of wives/partners.

Gentleman-like behaviour

Building rapport struck me as particularly easy across all interviews. This was even the case where there had been very little time between recruiting the participants and interviewing them, as was the case in the self-care study, where two men had agreed to take part in the study provided they could be interviewed straight away (in a consultation room of the pharmacy). Interviewees showed an eagerness to get the interview started and curiosity in what it would entail. Participants from the self-care and concordance study seemed both puzzled and pleased that I should have taken the effort to visit them at home and spend time with them asking questions. Participants from the early retirement study, who had had much more time and communication between learning about the study, agreeing to take part and finally meeting me for the interview, were explicit in how they looked forward to talking to me about their retirement experience. In this latter group, participants were spread across England, which involved travelling by train to interview them either at home or another arranged location (e.g. office of a charity for which one interviewee volunteered). On many of these occasions, participants offered to pick me up by car from the station and to drop me off again after the interview. On one occasion this involved opening and closing the passenger door for me. At the interview location they would offer tea or coffee. This might seem of little significance, and indeed both men and women interviewees did this. However, the men seemed to make a statement about it. Rarely did they just offer the hot beverage, but also an explanation that, although they were not good at working in the kitchen, coffee making was something they were proficient in. On other occasions the participants merely showed how good they were at being a host. Some also provided biscuits and showed pride in having thought about this extra offering. One of the early retirees confessed that his wife had reminded him in the morning to think about what he was going to offer me at our afternoon interview. This very hospitable behaviour was more pronounced in situations where the men were on their own at home, either because their wife/partner was out or because they lived on their own. Where wives were at home, the hospitable role was taken on by them.

Schwalbe and Wolkomir (2003), in their excellent chapter on interviewing men (though their focus is on young or middle-aged men) discuss the way interview settings can be experienced as potentially threatening to men or their masculine self, as the role of the interviewee appears to imply the relinquishing of control: 'To open oneself to interrogation is to put oneself in a vulnerable position, and thus to put one's masculinity further at risk' (ibid. p. 59). The authors argue that certain ways of phrasing questions allows the power relationship to be rebalanced, either towards the interviewee who might feel out of control (e.g. when being asked emotional questions) or the interviewer who can lose control (e.g. when a male participant becomes evasive). There seemed little to suggest that any of the older men I interviewed feared for the masculine self in this way. There were moments in several early retirement interviews where participants signalled they were not willing to explore a certain area further.[1] But issues around control or any potential threat to relinquishing control seemed to be outweighed by gentleman-like behaviour and a genuine interest in taking part in the research. This interest was also driven by seeing the interview as an opportunity to tell a story and to display knowledge and life experience. This level of generosity, that included insights into personal and emotional worlds, seemed possible because the men did not feel they had to compete with me. I believe that this was due to being a woman (and the gendered roles the interviewees associated with that, so that I was perceived as non-threatening), and as younger (i.e. someone who had not yet achieved what they had). These elements of age and achievement will be looked at more closely in the next section on generativity, whereas gender will be revisited in 'fatherly interest'.

Generativity

Generativity emerged from Erikson's work about psychological development stages (1950) and describes a stage in later life in which a concern of establishing and guiding the next generation as a way of contributing to society and helping to guide future generations (be this through raising a family or working toward the betterment of society) is considered to be important. Generativity is contrasted with stagnation that may affect people who are self-centred and unable or unwilling to help society move forward. It is not within the scope of this chapter to critique the normative concept of generativity but it struck me as a useful tool to understand why older male interviewees were keen to answer questions and elaborated when probed. For the early retirees the interview provided an opportunity to tell me about their successes: their career achievements, the financial and structural possibility of taking retirement early, the way in which they were living their retirement. The latter involving luxury as much as care for others, be this family or through voluntary work. However, the comparatively less well-off men from the self-care and concordance studies also used this opportunity to show how knowledgeable they were about their health and medical matters, for example, how they kept themselves fit, how they made use of primary health care, and how they made sense of conflicting medical advice. In parting with this knowledge and these ideas, the interviewees made sure they were passed on to me as a representative of a younger generation, and to me as a researcher who was going to present these ideas to a wider audience. As is known in qualitative interviewing, participants do not merely provide a description of events which have taken place in their lives. Instead, interviews provide an opportunity to spin a narrative about how certain things came about and the meaning they have. Schwalbe and Wolkomir (2003) warn that men can exaggerate rationality, autonomy and control as part of signifying a masculine self. And this would ring true with what I have described so far. However, the interviews I conducted were also full of accounts that showed the concern the men had for future generations and particularly their children, the financial worries they had had in earlier years, and the dependencies such situations had created in some cases. Moreover, the men spoke openly about poor health and difficult family relationships. The concept of generativity allows these narratives to be seen as important messages that the men wanted to pass on. Indeed interviewees talked about the desire to pass on knowledge to younger generations (often in terms of professional knowledge) but found that there was little demand. The qualitative interview provided an opportunity to share some knowledge and may account for the openness of interviewees.

The age mismatch also allowed me to ask questions about sensitive topics, such as health and financial matters that might not have been raised between an interviewer and interviewee similar in age and sex. Black and Rubinstein (2009), in conducting interviews with older African American men, describe a similar mismatch:

> As a white woman, the first author approached each respondent anxious to learn from his life and narrative, recognizing each man as an expert on the 'truth' of his life. Each respondent approached me as though he were educating me about the story of his life and issues discussed (ibid., p. 298).

This and my experience suggest that both sides of the interview avoided conceptual blindness, as neither assumed an understanding of any topics discussed. Viewing the interview settings through this generativity lens has made me conclude that the fact that I was significantly younger than the interviewees, allowed me to collect this rich data.

Fatherly interest

It was not uncommon for interviewees to ask me about my work and professional role either before or after the interview. On several occasions interviewees assumed that I did not have a professional role yet and that I was a student collecting data for a larger project. An extract from my interview notes illustrate this:

> We had arranged to meet at the station as he felt it would be easier to go by car than explain when to get off the bus. It was about a three-minute drive and we talked about [town] and he asked whether this was a postgraduate study of mine. He was rather hooked on the idea that I was still a student (e.g. he asked a little later whether I could use a student rail card for all the interview travelling I did). (Interview notes, interview 17, 31/10/2006)

During the same interview the interviewee kept comparing me to his daughter who was doing a degree. Although somewhat amusing in retrospect I remember feeling irritated about it at the time especially as I was not able to respond during the interview. At a personal level this felt patronising. Luckily, the interviewee was otherwise pleasant to talk to, so that it was not difficult to conduct the interview. Arising negative feelings for the interviewer resonate with what Schwalbe and Wolkomir (2003, p. 59) describe as 'the struggle for control', although the authors describe situations where interviewees deliberately test the ability of the interviewer to maintain control. However, feeling irritated or losing concentration can be detrimental to the quality of the interview. Thus in this instance, I did have concern about how the interviewee's view of me might influence what he said, such that if he thought that I was a student he would provide less depth than if he knew I had a professional role.

Interviewees with adult children were more interested in what I was doing and drew comparisons between me and their children. Another interviewee was particularly delighted when he heard about my recent lectureship and recognised the difficulties in establishing a career today in comparison to his own stable career path. In conversations about the research topic some interviewees across all three studies suggested authors I should read or other lines of enquiry that could be followed. Thus fatherly interest could take the shape of benign interest, to a shift in control. How these dynamics play out in the interview can have an impact on how sympathetic interviewer and interviewee can be towards each other, despite the professional stance qualitative researchers should have. Tang (2002) reports a similar experience in terms of 'motherly interest' in interviews with female professors as a doctoral student. However, fatherly interest can also be seen as an aspect of generativity, as it shows interest in a younger generation and a way of leaving a mark. As such, it can be used to enhance the rapport between interviewer and interviewee, and as a probe into topics such as children and future plans.

The role of wives/partners

Wives or partners emerged at different levels of the interviews. Interviewees would talk about their wife or partner during the interview and it was possible to ask about them. As the section on gentleman-like behaviour indicated, an interviewee's wife could also make her presence felt in terms of hospitality. Across all three studies many interviews were conducted in the participants' homes, and sometimes their wives or partners were also at home. Often I was only aware of their presence, because the interviewee told me, or because there had been a short moment of contact when entering the house. But I experienced three interviews, one in the self-care study and two in the early retirement study, where wives/partners made sure they were involved in the interview. On one occasion the interviewee's partner joined the room and prompted the interviewee when he did not know the name of a medication or certain treatment style. She also left before the interview ended. On another occasion the interviewee's wife had supplied us with tea and cake and then left the room. Just as the interview was coming to an end she reappeared indicating that this had been a long interview and that she was interested in contributing to it. This was very interesting as the interviewee had said little about this wife. I was then able to learn that she had found her retirement from being a maths teacher very difficult, and that she was now working in adult education on a voluntary basis. She also showed me photographs and pictures they had brought back from their extensive travels since retiring. A more challenging experience involved an interviewee's wife who joined the interview a short while after it had started and stayed its entire duration. My field notes from the interview describe the situation:

> The interview took place in the participant's living room and his wife served tea and a very large selection of biscuits and cakes, by which time the interview had already started. It was only after she had poured the tea, arranged plates and had then walked to the door that I realised that she had no intention of leaving the room. The situation was difficult for me to address and finally it was too late to ask her to leave, which would have also been quite rude. She didn't say anything until the end when I brought her into the conversation and asked her about her working life (he had mentioned her work several times without giving any details). (Interview notes, interview 20, 3/1/2007)

The interviewee gave a very detailed account of his work and retirement life, but also acted oddly oblivious to his wife's presence, which, if nothing else, gave some possible insight into their relationship. This experience has taught me to be more proactive in interview settings and to take control of events as they unravel. Bytheway (2011) talks of real life events and how they can enrich the data collected. If it is unavoidable to have a wife or partner joining the interview then it makes more sense, to understanding how the data was constructed, if she is allowed to contribute to the interview.

On reading this section one could also get the impression that the wives and partners who did make their presence felt, did so for more reasons than hospitality, namely: curiosity about the study, something their husband/partner had mentioned about the forthcoming interview, and not least curiosity about who was conducting the interview. Participants did not know my age before meeting me for the interview, but they knew my name. Thus, these different ways of interfering may have been their way of gaining an idea about to whom their husband/partner was talking to in such length and personal detail. Russell (2007), too, acknowledges the role of wives in the research she undertook with older men. She describes them as 'social organisers' (p. 183) who took charge of arranging the interview date, or whose husbands put them on the phone to clarify details of the project. All of the men I interviewed across the three studies arranged the interview date and venue directly with me but they did talk about negotiating diaries with their wives or partners if living in a relationship. These aspects suggest that wives and partners need to be taken into consideration when interviewing older men, including contingency plans of including them in the interview if necessary, or indeed designing the study in such a way that interviews are conducted jointly or separately with both.

Interviewing older men – Some conclusions

By combining the insights from the literature with the reflection on conducting interviews with older men I would like to make the case for a mismatch in terms of age and sex between interviewer and interviewee. The reason for this appears to be in the gendered roles of both the interviewer and interviewee, and in the nature of qualitative interviewing. Qualitative interviewing requires the interviewer to be empathic, listening and curious. These are traits that continue to be associated with gendered ideas of femininity, which are likely to be recognised and sought by older men when confronted with a young female interviewer. In addition, older men's masculinities, whether they are more traditional (e.g. fatherly, gentleman-like) or more contemporary (e.g. expressing emotions), can be receptive to the skills of an interviewer who encourages openness at the same time as protecting the interviewee's sense of self. This is not to say that a young female researcher will by default elicit rich data. Rather, this argument hopes to lift some of the concerns there are about using young researchers (whether male or female) in interviews with older people. An understanding of how rich the collected data is and what it might be missing, as well as potential biases, calls for reflexivity on the part of the researcher (Biggs, 2005).

However, some caution is due as the majority of men I interviewed were well off and could look back on a successful life. Whether the dynamics described above would play out as favourably in interviews with older men who are poorer or sicker needs to be established (Wenger, 2002). Issues of shame or embarrassment could mean that older men are less willing to open up to a younger woman. The interviews from the concordance and self-care study go some way in showing that older men who were less affluent and powerful did offer full accounts. However, more reflexive research is required in this area.

Finally, researchers and interviewers should acknowledge the role of gender in the relationship of interviewer and interviewee and the impact changing gender roles may have on conducting interviews in future. More research into how notions of masculinities develop and change across the lifecourse is needed and will allow qualitative researchers to develop their interview styles to enable openness, and empathy and avoid conceptual blindness.

Note

1   In one of the early retirement interviews the interviewee said he did not want to discuss his son's more difficult circumstances, but he did say that he supported him financially.

Annotated reading list

Biggs, S. (2005) 'Beyond appearances: Perspectives on identity in later life and some implications for method', *Journal of Gerontology: Social Sciences*, 60B (3) S118–S128.

This article tackles two topics: firstly, it examines contemporary debates on issues of identity in later life; secondly, it looks at methods, where Biggs encourages researcher reflexivity and takes a critical stance on current research practice in terms of interviewer bias.

Hearn, J. (1995) 'Imaging the aging of men' in M. Featherstone & A. Wernick (eds) *Images of Aging – Cultural Representations of Later Life*, pp. 97–115 (London: Routledge).

Hearn provides an analysis of the social and cultural construction of men in later life. By using examples from academic and autobiographical texts, films, advertising and magazines he shows how men in later life and their experiences have been viewed through varying cultural lenses, thereby conveying different meanings across the masculinity spectrum.

van den Hoonaard, K. (ed.)(2007) 'Special issue on aging and masculinity', *Journal of Aging Studies*, 21 (4) 277–368.

This special issue contains some of the most recent work on ageing men and masculinity. The nine papers from international authors cover a range of topics and methods, both qualitative and quantitative.

Schwalbe, M. & Wolkomir, M. (2003) 'Interviewing men' in J. Gubrium & J. Holstein (eds) *Inside Interviewing: New Lenses, New Concerns* (London: Sage).

Although this chapter is not concerned with interviewing older men, it provides an insightful analysis of the role of masculinity in interviewing men. Moreover it offers a plethora of practical advice on how to conduct interviews with men, for instance how to word questions that might rebalance the power-relationship of the interview, or how to illicit a more in-depth account from an interviewee who has been reticent.

References

Arber, S. & Ginn, J. (1991) *Gender and Later Life: A Sociological Analysis of Resources and Constraints* (London: Sage).

Asquith, N. (2009) 'Positive ageing, neoliberalism and Australian sociology', *Journal of Sociology*, 45 (3) 255–69.

Bennett, K. (2007) '"No sissy stuff": Towards a theory of masculinity and emotional expression in older widowed men', *Journal of Aging Studies*, 21, 347–56.

Biggs, S. (2005) 'Beyond appearances: Perspectives on identity in later life and some implications for method', *Journal of Gerontology: Social Sciences*, 60B (3), S118–S128.

Black, H. & Rubinstein, R. (2009) 'The effect of suffering on generativity: Accounts of elderly African American men', *Journals of Gerontology, Series B: Psychological Sciences and Social Sciences*, 64B (2), 296–303.

Bytheway, B. (2011) *Unmasking Age – The Significance of Age in Social Research* (London: Policy Press).

Calasanti, T. (2004) 'Feminist gerontology and old men', *Journals of Gerontology, Series B: Social Sciences*, 59B (6) S305–S314.

Davidson, K. & Meadows, R. (2010) 'Older men's health: The role of marital status and masculinity' in B. Gough & S. Robertson (eds) *Men, Masculinity and Health* (Basingstoke: Palgrave Macmillan).

Erikson, E. (1950) [1995] *Childhood and Society* (London: Vintage).

Gilleard, C. & Higgs, P. (2000) *Cultures of Ageing* (Harlow: Prentice Hall).

Gilleard, C. & Higgs, P. (2005) *Contexts of Ageing: Class, Cohort and Community* (Cambridge: Polity).

Gott, M. & Hinchliff, S. (2003) 'How important is sex in later life? The views of older people', *Social Science and Medicine*, 56 (8) 1617–28.

Hammersley, M. (2008) *Questioning Qualitative Inquiry: Critical Essays* (London: Sage).

Hearn, J. (1990) 'Series Editor's Preface' in D. Jackson, *Unmasking Masculinity: A Critical Autobiography* (London: Unwin Hyman).

Hearn, J. (1995) 'Imaging the aging of men' in M. Featherstone & A. Wernick (eds) *Images of Aging – Cultural Representations of Later Life*, pp. 97–115 (London: Routledge).

Hearn, J. & Pringle, K. (2006) 'Studying men in Europe' in J. Hearn & K. Pringle (eds) *European Perspectives on Men and Masculinities. National and Transnational Approaches* (Basingstoke: Palgrave Macmillan).

Holstein, J. & Gubrium, J. (2003) 'Introduction – Inside interviewing: New lenses, new concerns' in J. Holstein & J. Gubrium (eds) *Inside Interviewing: New Lenses, New Concerns* (London: Sage).

Hurd Clark, L., Griffin, M. & The PACC Research Team (2008) 'Failing bodies: Body image and multiple chronic conditions in later life', *Qualitative Health Research*, 18 (8) 1084–95.

Johnson, J. (2001) 'In-depth Interviewing' in J. Gubrium & J. Holstein (eds) *Handbook of Interview Research* (Thousand Oak: Sage).

Jones, I.R., Leontowitsch, M. and Higgs, P. (2010) 'The experience of retirement in second modernity: Generational habitus among retired senior managers', *Sociology*, 44 (1) 102–20.

Katz, S. & Marshall, B. (2003) 'New sex for old life: Lifestyle, consumerism, and the ethics of aging well', *Journal of Aging Studies*, 17 (1) 3–16.

Laslett, P. (1989) *A Fresh Map of Life: The Emergence of the Third Age* (London: Weidenfeld and Nicolson).

Lee, A. (2008) 'Finding the way to the end of the rainbow: A researcher's insight investigating British older gay men's lives', *Sociological Research Online*, 13 (1) 6.

Leontowitsch, M., Higgs, P., Stevenson, F. & Jones, I.R. (2010) 'Taking care of yourself in later life: A qualitative study into the use of non-prescription medicines by people aged 60+', *Health*, 14 (2) 213–31.

Leontowitsch, M., Stevenson, F., Nazareth, I. & Duggan, C. (2005) '"At the moment it is just of couple of eccentrics doing it": Concordance in day-to-day practice', *International Journal of Pharmacy Practice*, 13 (4) 273–80.

Mann, R. (2007) 'Out of the shadows?: Grandfatherhood and masculinities', *Journal of Aging Studies*, 21 (4) 281–91.

ONS (Office of National Statistics) (2010) *Life Expectancy*, http://www.statistics.gov.uk/cci/nugget.asp?id=168, date accessed 27 July 2011.

Phoenix, C. & Smith, B. (2011) 'Telling a (good?) counterstory of aging: Natural bodybuilding meets the narrative of decline', *Journals of Gerontology, Series B*, 66 (5) 628–39.

Phoenix, C. & Sparkes, A. (2009) 'Being Fred: Big stories and the accomplishment of a positive ageing identity', *Qualitative Research*, 9 (2) 219–36.

Potts, A., Grace, V., Vares, T. & Gavey, N. (2006) '"Sex for life"? Men's counterstories on "erectile dysfunction", male sexuality and ageing', *Sociology of Health and Illness*, 28 (3) 306–29.

Reich, J. (2007) 'Unpacking the "pimp case": Aging black masculinity and grandchild placement in the Child Welfare System', *Journal of Aging Studies*, 21 (4) 292–301.

Robeiro, O., Paúl, C. & Nogueira, C. (2007) 'Real men, real husbands: Caregiving and masculinities in later life', *Journal of Aging Studies*, 21 (4) 302–13.

Russell, C. (2007) 'What do older women and men want? Gender differences in the "lived experience" of ageing', *Current Sociology*, 55 (2) 173–92.

Russell, C. & Porter, M. (2003) 'Single men in disadvantaged households: Narratives of meaning around everyday life', *Ageing International*, 28 (4) 359–71.

Schwalbe, M. & Wolkomir, M. (2003) 'Interviewing men' in J. Gubrium & J. Holstein (eds) *Inside Interviewing: New Lenses, New Concerns* (London: Sage).

Silver, C. (2003) 'Gendered identities in old age: Towards (de)gendering?', *Journal of Aging Studies*, 17 (4) 379–97.

Silverman, D. (1993) *Interpreting Qualitative Data* (London: Sage).

Smith, J., Braunack-Mayer, A., Wittert, G. & Warin, M. (2007) '"I've been independent for so damn long!": Independence, masculinity and aging in a help-seeking context', *Journal of Aging Studies*, 21 (4) 325–35.

Stephenson, P., Wolfe, N. & Coughlan, R. (1999) 'A methodological discourse on gender, independence, and frailty: Applied dimensions of identity construction in old age', *Journal of Aging Studies*, 13 (4) 391–401.

Suen, Y.T. (2010) 'Reflecting on studying older men's lives: Thinking across boundaries' in M. Harrison & P. Ward (eds) *Beyond Borders: Masculinities and Margins*, pp. 198–203 (Tennessee: Men's Studies Press).

Suen, Y.T. (2011) 'Men on their own: How do older men experience and negotiate singlehood?' Paper presented at the British Sociological Association Annual Conference, London.

Tang, N. (2002) 'Interviewer and interviewee relationships between women', *Sociology*, 36 (3) 703–21.

Thompson, E. (1994) 'Older men as invisible men in contemporary society' in E. Thompson (ed.) *Older Men's Lives* (London: Sage).

Thompson, E. (2006) 'Images of old men's masculinity: Still a man?', *Sex Roles*, 55, 633–48.

Tulle, E. (2008) *Ageing, the Body and Social Change: Running in Later Life* (Basingstoke: Palgrave Macmillan).

van den Hoonaard, K. (2007) 'Aging and masculinity: A topic whose time has come', *Journal of Aging Studies*, 21, 277–80.

van den Hoonaard, K. (2010) *By Himself: The Older Man's Experience of Widowhood* (Toronto: University of Toronto Press).

Wenger, C. (2002) 'Interviewing older people' in J. Gubrium & J. Holstein (eds) *Inside Interviewing: New Lenses, New Concerns* (London: Sage).

Part III

Old and New Qualitative Methods

7
Using Focus Groups for Researching End-of-Life Care Issues with Older People

Jane Seymour

Meeting the needs of older people as they approach the end of their lives is a key public health challenge as changes in demography and epidemiology transform trends in longevity, morbidity and mortality. These changes mean that death and chronic illness have given rise to palliative care needs that are largely concentrated in older age groups (Davies & Higginson, 2004). Yet it has long been recognised that health and social care services lack responsiveness to the needs of older people in the last year of their lives (Seymour et al., 2005). Current policy direction reflects recognition that palliative and end-of-life care is not only about the management of the last days of life but also about supporting people with long-term health conditions to live as comfortably as possible until they die. A key emphasis is on enabling participation in care planning both during and in advance of any illness, and taking into account the individual's values and preferences. In England, a National End-of-Life Care Strategy (DoH, 2008) advocates the development of approaches to raising public awareness about end-of-life care issues so that when people face a final illness they may better articulate their needs for care. Internationally it is recognised that finding new ways of engaging and understanding the views, experiences and perspectives of potential users of palliative and end-of-life care services is an important area of research, as well as policy and practice development. In doing so it addresses the critical public health issue of enabling older people to have a better quality of life as they move towards death (WHO Europe, 2011). Sociologically informed research addressing the management and social construction of dying, and of the 'good' death has been published in the field of palliative care studies (Clark & Seymour, 1999), while gerontological research has largely focused on processes of ageing. However, it is only comparatively recently, with one or two seminal exceptions (e.g. Williams, 1990), that a body of work has begun to emerge which explores the intersection of dying and ageing (e.g. Aldred et al., 2005; Pleschberger, 2007). As a result, there is an increasing emphasis on using qualitative research among older people to gain their perspectives and experiences of palliative and end-of-life care, but comparatively few reports about the methodological and ethical issues involved (Kendall et al., 2007).

This chapter draws on experiences of community-based studies conducted by colleagues and myself over the last decade. It considers, based on these experiences, the contribution of focus groups to the pursuit

Table 7.1   Overview of Key Publications and Studies

Study reference	Aims	Methods/participants
Exploring the impact of sexual orientation on experiences and concerns about end-of-life care and on bereavement for lesbian, gay and bisexual elders (Almack et al., 2010).	To explore the impact of sexual orientation on experiences and concerns about end-of-life care and on bereavement within same sex relationships.	Four focus groups with participants in a lesbian, gay and bisexual (LGB) support network: older men (n=10) and women (n=5). Part of a larger study involving older adults and carers from five community groups.
'At the foot of a very long ladder': discussing the end-of-life with older people and informal caregivers (Clarke & Seymour, 2010).	To understand the concerns older people may have about end-of-life issues and to provide advice and information to help them to address these.	Focus groups with 74 older people and family carers, recruited via a national charity: Help the Aged, from community groups in England.
Older people's views about home as a place of care at the end of life (Gott et al., 2004). (See also, Seymour et al., 2004, Seymour et al., 2005).	To examine older peoples' beliefs and risk perceptions regarding the use of innovative health technologies in end-of-life care and to relate these beliefs to their ideas about 'natural death'.	Six focus groups and 39 interviews with 77 older people living in three socio-demographically contrasting community areas in one city.
Hospice or Home? Expectations about end-of-life care among older white and Chinese people living in the UK (Seymour et al., 2007).	To compare the perspectives and experiences of older white and Chinese people living in England about places of care and death.	Secondary analysis of interview and focus group data from studies of older 'white' and 'Chinese' elders, living in three cities in England.

of end-of-life care research among older people, and offers a reflective account of how issues associated with recruitment, informed consent and facilitating the discussion of potentially distressing material in a group setting may be managed. The chapter concludes by considering the limitations of focus groups as a methodology for researching end-of-life issues with older people, and offers a set of recommendations for planning and conducting such research sensitively and with due regard for the range and heterogeneity of needs for support that older people who take part in the latter may have. Table 7.1 provides an overview of studies using focus groups, and associated key references, from which these reflections are drawn.

The use of focus groups in social and health-related research

Focus groups possess elements of participant observation and individual interviews (Madriz, 2000), and are primarily a means of listening to views on a specific area of inquiry within a non-threatening environment (Morgan, 1997). Focus groups are a popular method for accessing understandings of illness and for examining people's experiences of health and health services (Duncan & Morgan, 1994) and have been advocated as a useful means of engaging with users in developing health care policy and practice (Thornton, 1996; Owen, 2001). The origins of focus groups are located frequently in the opinion gathering activities of market or political research, where speed, low cost and flexibility often take precedence over any concerns with representativeness or rigour. However, as Reed and Payton (1997) point out, focus groups have also long been part of a range of techniques employed in the anthropological and ethnographic tradition of qualitative social science. In this tradition, focus groups are used explicitly because of their potential to allow study of the role of social interaction and unfolding conversation in opinion formation, especially where the subject of study is little understood and talked about infrequently in day-to-day life. Thus, the concern is less directed towards uncovering stable 'facts' about opinions and attitudes that may be elicited from the different participants in the focus groups, and more to do with:

> ...the process of developing a group perspective or position among a particular set of people...[although] people come to a focus group with particular ideas and [views] (Reed & Payton, 1997, p. 770).

As Kitzinger (1995) suggests, focus groups are an important opportunity for examining the way in which perspectives are negotiated and developed between their members:

> Focus groups are a form of group interview that capitalises on communication between research participants in order to generate data...This means that instead of the researcher asking each person to respond to a question in turn, people are encouraged to talk to one another: asking questions, exchanging anecdotes and commenting on each other's experiences and points of view (Kitzinger, 1995, p. 299).

Focus groups are, then, a potentially successful means of exploring sensitive issues such as death and dying. Group processes may help people to explore such issues in their own terms and aid them to generate questions and priorities that relate to the issues under study and which can be then further explored during the research process. As Kitzinger notes:

> Group work can actively facilitate the discussion of taboo topics because the less inhibited members of the group break the ice for the shyer participants. Participants can also provide mutual support in expressing feelings that are common to the group but which they consider to deviate from mainstream culture, or the assumed culture of the researcher (Kitzinger, 1995, p. 300).

Where groups are 'naturally occurring' (Morgan, 1997), i.e. when members are already known to each other by dint of some allegiance or membership of an established organisation, then focus groups can serve the dual purpose of providing a comfortable and supportive environment for participants, and of introducing the researcher to important aspects of the cultural values that underpin the wider organisation. In this way, focus groups, when carefully conducted, provide an important means of accessing groups who may be otherwise relatively marginalised in research, and of 'opening up' for discussion sensitive topics that may be little understood and infrequently talked about.

Conducting focus groups with older people: Some general issues

Despite their potential as a research method, there are few reports of any special considerations that may be required when trying to involve older and/or disabled participants in focus group studies in the fields of medicine and health care. However, studies in areas such as education (Keller et al., 1987) and nutrition (Crockett et al., 1990) have shown that focus groups with older people can be successful, providing that steps are taken to ameliorate the sensory, physical and mental impairments associated with older age. It also seems that success may be determined, even more critically than usual, by the overall study design, organisation of the questions, and moderator skills (Morgan, 1997). Owen (2001), reporting on the use of focus groups with potentially vulnerable older people, emphasises the time-consuming nature of this type of research, together with the importance of experienced facilitators who can manage the consequences of personal disclosure, and provide mechanisms of support to research participants where these are necessary. Providing a rare and detailed account of the plethora of practical and interpersonal considerations to be taken account of when using focus groups with older people, Warren et al. (2003) report in a Sheffield study, on experiences of trying to maximise the involvement of older women from a number of diverse and often deprived or marginalised communities. In *Older Women's Lives and Voices* (Warren et al., 2003) the aim was to find out about the experiences of women aged over 50 in using public services. It built on some pilot work involving a small number of older women from three ethnic groups. This pilot project highlighted some important factors that informed the design and conduct of the larger project, most notably that:

> ...one of the biggest problems lay in communication. Their own illiteracy, combined with the general failure of the general public service sector to provide adequate translators and advocates compounded the negative aspects of these factors. Quite often it was the case that women were speaking out loud and clear but policy makers, service providers – their own families even – were simply not hearing or listening (Warren et al., 2003, p. 25).

In Warren, Cook and Clarke's follow-on project, 100 older women, drawn from Black-Caribbean, Irish, Chinese, Somali and White British communities, were involved in a series of discussion groups focusing around 'growing older', 'using services' and 'having a say'. Warren et al. (2003) describe recruitment to these groups as both 'pragmatic and serendipitous'. It involved intensive networking, both through formal and informal channels, and an extremely time-consuming process of negotiating the fine detail of the involvement of the groups in the research. Once negotiations were complete, a new series of issues had to be addressed, as Warren et al. reflect:

> Working with older women with various degrees of disability, from 11 different ethnic groups and at the community level, required the provision of appropriate and suitably equipped venues, transport, refreshments, and translations services, underpinned by flexible agendas, a constant flow of phone calls and a raft of written information, ...groups were allocated funds directly to arrange refreshments that suited their tastes...[and in relation to participants with hearing difficulties] efforts to accommodate their needs included: sitting in a circle so that everyone had a clear view of each other; asking participants to speak clearly and give everyone a turn in the conversation; providing individual scribes to take notes; and summarizing points on flip charts. Typically, however, efforts were not sustained for the whole meeting and the women with hearing problems either sat in (frustrated) silence or, conversely, attempted to dominate the conversation (Warren et al., 2003, p. 26).

As Warren, Cook and Clarke highlight so clearly, the success or failure of focus groups in studies involving older people may be determined by the ability of researchers to take account of frailty, illness, disability or cognitive impairment. However, in research which invites older people to consider issues about end-of-life care (whether hypothetically or from experience), an additional set of issues, associated with perceived sensitivity of the subjects of death and dying in contemporary society, often arises. In addition, the problems of gaining access to older people (especially those in very late old age or in particular types of care setting) are multiplied several times in end-of-life care research (e.g. Hall et al., 2009). Cultural and language differences when trying to include older people from minority ethnic groups often further complicate matters, not only in relation to access and recruitment, but also in relation to trying to gain informed consent. Here cultural sensitivities may make the whole enterprise appear highly risky and unsafe (Hughes et al., 1995).

Reflections on involving older people in focus groups in end-of-life care research

In planning our first studies examining older adults' perceptions, attitudes and experiences of end-of-life care issues in the early 2000s, valuable lessons were available from reports of focus group studies conducted in the United States of America (USA). These explored the meaning of 'good death' with patients, families and service providers in the US (Steinhauser et al., 2000), and end-of-life decision-making with women from older age and minority ethnic groups (Morrow, 1997). These experiences confirmed the need for small group size, careful laying down of ground rules, and attention to clarity – alongside the more obvious issues of needing at least two researchers with experience of sensitive research issues for group facilitation, and the use of carefully worded probing questions. However, before focus groups could take place, we (my colleagues and I) had to work with gatekeepers to access individuals, explain what taking part in the studies would involve, and gain informed consent. These preliminary aspects are explored below.

Gaining access

In the studies outlined in Table 7.1 we used a similar methodology to prepare the ground and try to ensure adequate levels of participation and representation. We have always worked via community groups using the principles of purposive or snowballing sampling, finding that when focus groups take place, this means that older participants are more likely to feel comfortable with their peers and to be familiar with the discussion venues. The one exception has been a study commissioned by the United Kingdom (UK) charity formerly known as Help the Aged (now Age UK), in which we employed a range of recruitment methods including advertisements in the local press/community group magazines; and letters/emails to representatives of older people groups and known contacts (Clarke & Seymour, 2010). With this one exception, as a general rule once we identified potentially interesting groups, we wrote to the relevant group contact person, enclosing brief details of the particular study. We then arranged a face-to-face meeting with the group representative. These face-to-face meetings are important, since they give people who effectively control access to the group membership an opportunity to 'vet' the study, our proposed conduct and our trustworthiness. Moreover, they provide an opportunity to discuss issues such as the best venue and other practical aspects. We kept inclusion and exclusion criteria flexible, explaining how we are interested in recruiting older men and women who feel able to discuss end-of-life care with a small group of people with whom they are familiar, together with two researchers. When we first embarked on this sort of research we tended to organise single focus groups; latterly, we have refined an approach reported in Clarke and Seymour (2010) that involves workshops comprising a number of parallel discussion groups (see below for more detail). In all of the studies, we have taken a loose approach to the definition of 'older', relying on self-definitions of potential focus group members.

Inviting participation and gaining informed consent

Before focus groups were convened, working with gatekeepers usually resulted in a preliminary meeting with potential participants at their regular community group. In most cases working via gatekeepers in this manner enabled access to a diverse range of older people and was helpful in the process of communicating to them the subject of our research. However, there have been occasions where, although we have been invited to meet potential participants, it became clear that the gatekeepers had said little or nothing about the focus of our research, or indeed about our motivations in requesting the meeting. This was illustrated well by a study of Chinese elders living in the UK (Seymour et al., 2007) where a large attendance was gained at a preliminary 'pre-participation' meeting convened by a community leader. It became obvious that attendees were expecting a 'health talk' from our very able Chinese researcher, and many were irritated that this expectation was not fulfilled.

Approaching preliminary meetings in an honest and straightforward manner has been important, although we have found this to be a delicate process where the gatekeeper feels in any way uncomfortable or particularly sensitive about the topics. A valuable way of proceeding has been to introduce the idea of taking part in research and then to say a little about our backgrounds as palliative and end-of-life care researchers, and discuss with people what these terms mean to them. We have used a number of approaches to try to ensure that potential participants are as well informed as possible about the aims and objectives of our study, what exactly taking part will involve (perhaps by sharing some broad themes or questions), the potential risks of taking part (here we talk about the risk of recalling experiences of bereavement or upsetting experiences of care), and what we will do with the information. It has been of crucial importance to convey the key point that we are not assuming that serious illness and dying is something which inevitably and only happens to older people. We have instead emphasised the various problems in accessing the views of older people about serious illness and dying, and in exploring the special difficulties faced by some older people.

We have made use of a combination of written, verbal and pictorial information, since older people differ in their ability to read and hear our information. In one of our later studies, we developed an information booklet co-written with older people (Seymour et al., 2006) about end-of-life care issues, because of the sheer lack of information available. This has subsequently been a resource that we leave with people, whether or not they decide to take part in any focus groups. These booklets have been helpful both in addressing questions people may have about end-of-life care issues, and in enabling individuals to consider their own perspectives on these topics before their participating in any research-based activity. However, we have encountered individuals with primary illiteracy and others who cannot read because of sight loss, so this is not an answer to all problems of understanding and communication. On the whole, people invited in this way have seemed motivated enough to contribute to the studies, and in many cases appear to have given the issues much thought before taking part. Clearly, a disadvantage is the potential 'shaping' or constraining of views that provision of pre-participation information of this sort may elicit.

One of the methodological problems we encountered is that the very process of inviting people to take part in the project in a group context encouraged some to immediately discuss the issues in an impromptu, and often animated and engaged way, while others have tried to disengage themselves. In these situations we steered the conversation to a close while pointing out that the issues being aired are exactly the ground that would be covered in a focus group discussion:

> ...(we) asked whether one or two concrete examples would help, which the key spokesperson said would be helpful. I then explained about a personal story I had read the day or so before in the Alzheimer's Disease Society newsletter written by a man with the early stages of the disease and in which he referred to 'living wills' and 'advance directives'. I explained that we would use this sort of story in the focus group to kick-start the discussion. This led to immediate interest around the table with everyone discussing what they thought a living will was (some confusion over whether it was possible to buy one at the post office) and one woman announcing what she wouldn't want if she had a stroke (resuscitation) but at the same time saying that she would challenge any doctor to say she only had '48 hours to live'... We were becoming increasingly concerned that the focus group was almost taking place by default – I therefore said that it was time to draw a line under the very interesting discussion since we could not be sure that everyone was happy with it (one woman did, in particular, look very anxious). I explained that the discussion could be seen as an example of how the focus group might run, but

that we had to be sure that everyone who was at the focus group was participating willingly, and not as a captive audience (Field notes: preliminary meeting with participants from an Age Concern community group).

The constraints of ethical committee procedures means that even where everyone in a group appears to feel comparatively at ease with preliminary subject explorations, we have had to ask them to engage with the formalities of paperwork, so as to gain their written or at least verbal consent. In some cases, this deterred participation or posed challenges to us in managing the research exchanges. This was most obvious in a study that involved (as one aspect of a lager project) older adults living in an extra care housing complex (Seymour et al., 2009). We found at a preliminary meeting in the housing complex social area that many were interested in the focus of the study, and were happy to make comments, report experiences and engage in discussion about the issues at hand. One person reported that *'we think about this more than people realise'*, prompting others to recount stories of bereavement or of experiences of care (both good and bad) in hospitals and other care settings. However, in spite of this interest, most did not want to engage subsequently in the more formal process of participating in a focus group. On another occasion, when in the process of conducting a focus group at a community centre with participants from an Irish elders' group, we were holding the second of two focus group discussions in a room that led out into the main meeting room. A small group had given written consent, but one older man decided that our material was interesting and that maybe some of his friends in the main room would have comments to make. He threw open the double doors and invited people to come in. We ended up with people coming in and out, making contributions as they saw fit. It has been observed elsewhere that ethical conduct in such situations is not something that can always be pre-planned or strictly constrained by the plan presented for ethical committee review; rather, it is a matter of thinking on one's feet according to any particular situation (De Laine, 2000).

Managing focus group discussions of end-of-life care issues

Once a focus group discussion was arranged, we sought to find a balance in the focus groups, between the expression of individual views, and coverage of the range of issues in which we were interested, with the need to set some clear framework around the material. We tried to ensure that people could distance themselves from the material if they wished, while allowing space and time for personal stories if participants desired. Allowing people to maintain their sense of privacy, but also providing enough material so as to have a general discussion if preferred, has been crucial. In one of our early studies we successfully used the vignette technique (Finch, 1987), which has been used in written materials employed in focus groups about sensitive issues (Brondani et al., 2008). However, we adapted it so that the material could be shown on a series of simple PowerPoint™ slides. We thus developed an attractive but simple pictorial 'aide-memoire', in which the following key themes were presented

- Where is the best place to be cared for (home, hospital, nursing home or hospice)?
- Using technology to prolong life (resuscitation and artificial feeding)
- Using technology to give comfort (terminal sedation and morphine)
- Who should decide (Clinical staff/patient/family communications and advance care planning)?

There were some disadvantages to this approach. Because it is fairly structured it may not have created sufficient opportunities for participants to raise other issues of concern. However, on the whole, we believe that there were several advantages attached to presenting the issues in this way. Firstly, the computer equipment itself generated interest and discussion, and this was a useful ice-breaker. Secondly, the structured approach facilitated data analysis, and made it easier to use the material to inform the development of an interview aide-memoire in a subsequent study phase. Thirdly, the material gave enormous flexibility with which to vary the pace of the discussion, and to revisit or perhaps skip over particular images or words where appropriate. It also acted as a useful marker for the end of a topic: on the occasions when the discussion became deeply personal there was at times a need to make a break, and the slides made this easier. Fourthly, and perhaps most importantly, we realised, unexpectedly, that group participants viewed the slide show as something akin to watching television (hence the invitation from the participant in the Irish Elders' group to his companions, reported above). Television is something with which most people are comfortable and familiar, and many people sometimes watch programmes that include discussion of taboo or 'risky' material not readily discussed in day-to-day life.

The material shown generated a lot of interest, and people asked many questions that we tried to answer in a straightforward manner. The questions themselves proved very revealing, giving us an insight into the types of need that people have for information about end-of-life care.

Sometimes, stories were told that revealed a need for information and discussion and this ultimately prompted us to develop the information booklet described above (Seymour et al., 2006).

The following is perhaps an extreme example of the sort of 'difficult' story that emerged in one of the groups and which needed an immediate response through the discussion of the meaning of euthanasia and the differences between relieving suffering, and intention to kill:

> ...[x] told us that: 'I've done euthanasia'. She recounted a harrowing story of 18 years before, when she cared for her mother who died a lingering death from cancer. She had been left with morphine syrup to give her mother and next to no guidance as to its use (Field note: focus group).

At the end of each group we 'debriefed' participants informally, often over lunch, asking them what they thought of the way in which we handled the discussion and how they felt about it. We offered follow-up contact to discuss anything that has been considered in the group, or to answer any questions that might arise during subsequent days and weeks. We also provided the names and addresses of bereavement care organisations. Feedback has included the following:

> ...they told us that they thought we [the researchers] had handled the material and discussion 'very well' and 'sensitively', and that they would, because of this, be keen to be involved further. Overall they seemed surprised and relieved that we managed successfully a risky topic...We do not underestimate the importance of this exchange to the eventual success of the project... (Field note following focus group).

In later projects we drew upon all these lessons to run one-day workshops, which we sometimes called 'listening events' (Clarke & Seymour, 2010), in which up to 25 or 30 older people participated, together with gatekeepers or stakeholders from their community groups. These events, which were divided into morning and afternoon sessions, started with an introduction to the background and aims of the events, and time for questions. Participants were then invited to join small discussion groups, carefully facilitated by two experienced members of the research teams (and in some cases by co-facilitators drawn from groups of older people who worked with us as 'peer educators' (Sanders et al., 2006; Seymour et al., 2009; Clarke & Seymour, 2010). The open discussions held in the morning session gave participants the opportunity to relate their own experiences and concerns about death and dying, which both reflected the methods used by others who have accessed older people's views about perspectives, needs, and service provision (see for example, Barnes, 2005) and allows for the crucially important process of storytelling about loss. In the afternoon, structured discussions were facilitated using the booklet *Planning for Choice in End-of-Life Care* (Seymour et al., 2006), which employs vignettes to introduce a range of topics about end-of-life care, and was co-authored by five older people.

Interaction in the focus groups often involved participants asking questions, either of the researchers or of each other, and then reaching some shared understanding. This process is particularly evident where participants have little direct experience of the topic under discussion, or little previous opportunity to think through their opinions. The following is an example of a discussion which took place in the early 2000s about advance care planning and 'living wills'; subjects about which participants knew comparatively little even though we had provided written and pictorial information. The exchange took place before we started to use the more comprehensive information booklet, and prior to the introduction in England and Wales of the Mental Capacity Act (2005), which makes provision for the nomination of lasting power of attorney for health and welfare, and enables the completion of potentially legally binding advance decisions concerning the refusal of treatment:

> Researcher: Given that in some circumstances some people are aware that perhaps they have got an illness that is [progressive]; do you think there is a role for making an advance care decision? Is it something that you would do?
> Claire: I would
> Claudia: I think I would like to talk it over with the family, you know, I mean I've made the will and everything
> Claire: I've thought about it already
> Clarice: Yes, I've taken one out, a power of attorney and given it to my youngest daughter
> Researcher: What about in terms of treatment though, because in some places you can have a power of attorney for treatment
> Clarice: Exactly, I didn't realise that until I've just been reading this
> Researcher: It's not legal in England at the moment but it is legal in Scotland and the law may eventually change here; but in this country if you have given an instruction about care say, for example, if I become more ill with this particular type of illness I have got I would not wish

to be ventilated or I would wish not to be resuscitated, as long as they can be sure that you have properly anticipated the circumstances then I think the courts would take a very dim view if the doctor went against that. What do you think about that?

Claire: Would you necessarily, you'd sign it yourself, would you, or would someone else [be] signing it, say that was your signature or would it have to be the doctor?

Researcher: It can be a note written in your notes as a result of the conversation with the doctor

Claire: It would have to be someone like that rather than say a friend or anything like that?

Researcher: If it was something that you'd written down, then, and they could be sure that you had actually written that document that would be good enough

Claire: If you signed it and then someone else signed it to say that was your signature

Catherine: Like your will

Claudia: Like your will

Claire: Yeah, same principle as a will, so it really wants attaching to your will, well no because they wouldn't look at your will would they until you were dead would they?

Catherine: No don't attach it before time though

Researcher: I get the impression it's actually quite a new thing and you perhaps would like a little bit more time to think about it

Catherine: Not thought about that

Claudia: Yes but you've got to have somebody near to [talk to]

Catherine: I mean I've paid for my funeral but it's the bit before that wants seeing to isn't it! (Focus group transcript)

Limitations of using focus groups and key recommendations

One of the disadvantages of focus groups is that discussion may be dominated by vocal participants, or by those who have come with a very particular issue to report. In very practical terms, it can be difficult to achieve a balance between giving each participant time to tell their own story and ensuring that everyone who wants to speak feels able to do so. In addition, some participants may be reluctant to contradict others when the topic under discussion causes embarrassment or distress; alternatively, feelings may run high, with participants being openly critical or intolerant of others' views (Brondani et al., 2008). The relatively structured approach we adopted (especially in the early studies; less so in the later ones) in introducing the research issues meant that we risked reducing the opportunities available to participants to raise those issues of central personal concern. Furthermore, our purposive and snowballing sampling techniques will not have generated data that is representative, and indeed there may be something highly unusual about those who are willing to discuss death and dying in a group setting. When personal experiences have been discussed the small size and structure of the groups have often given participants comfort, empowerment and reassurance from the empathy of others. An example of where this was especially apparent was when one older man gathered courage in a workshop event to express what was obviously a deep-seated fear that he would die alone, describing this as signifying an 'undignified end' or perhaps lack of social worth:

> I've moved to a flat and there's nobody in the block. I worry about dying alone and suffering a stroke. If I took ill – what would happen? It's the indignity of dying alone.

He felt that media reports had 'triggered worrying about these things', although he had personal experience of such an incident:

> I attend a reading group and one member had been dead for five days before anyone found them (Focus group transcript).

Other group participants not only empathised with these worries, but shared in very practical terms ideas and information about how they had addressed the issues. It was clear later on in the day that this had been found to be extremely helpful by the person concerned. Of course we do not know how many fears may have been left unaddressed because the people concerned did not have the confidence to air them.

Along the way, my colleagues and I learnt some key principles of employing focus groups in studies of end-of-life care involving older people, and I will conclude by sharing these with readers. Many of these are common sense, but they are crucially important in studies of this kind:

- Work with gatekeepers to find out the preferred way of contacting potential participants and introducing your study topic.
- Providing pre-discussion group information when the topic is potentially sensitive can help 'break the ice', and the benefits in terms of

reducing anxiety seem to outweigh the risks of overly constraining the content of discussions.

- Keep the groups very small; mainstream focus group recommendations about size can sometimes be unmanageable when time is needed to recount personal stories without being hurried.
- Many people see the focus group as an opportunity to tell a story about loss. Allow plenty of time for discussions, and have at least two facilitators and an observer who is free to offer support if participants appear distressed. Have a space set on one side, where people can go to sit quietly in case of need.
- Often it is difficult to achieve a balance between participants feeling they have had adequate opportunity to speak about matters that are important to them (which may deviate from the topics under discussion), and of ensuring that everyone in the group feels able to have their say. Make attempts to address this by ensuring time to set ground rules. Explain that all opinions should be listened to and respected and that any one view or statement is not more desirable than another.
- If one person's views dominate, remind participants of the original ground rules of respect for individuals and their views, keeping to the topics discussed, not criticising others – even if they disagreed with their opinions – and not interrupting when someone was speaking. This is particularly pertinent for discussions of a sensitive nature which may raise powerful emotions.
- Make time for debriefing, not only with focus group participants, but also with the research team, and plan how this will be done in advance.
- Provide follow-up contact options and local bereavement support service information: not many people are likely to avail themselves of these links, but some do.
- Do try to go back to the focus group participants to tell them how you made sense of what they said to you: we have always found that people find this very helpful.

Acknowledgement

This chapter is based on a paper published in Ageing and Society as: Seymour, J.E., Bellamy, G., Gott, M., Ahmedzai, S. & Clark, D. (2002) 'Using focus groups to explore older people's attitudes to end-of-life care', *Ageing and Society*, 22, 517–26.

Annotated reading list

Brondani, M.A., MacEntee, M.I., Bryant, S.R. & O'Neill, B. (2008) 'Using written vignettes in focus groups among older adults to discuss oral health as a sensitive topic', *Qualitative Health Research*, 18, 1145–53.

This paper provides an interesting, insightful and very practical account of using vignettes to encourage older people to talk about their experiences with a particular health issue.

Owen, S. (2001) 'The practical, methodological and ethical dilemmas of conducting focus groups with vulnerable clients', *Journal of Advanced Nursing*, 36, 652–8.

This paper is a useful overview of the dilemmas that researchers may face in working with research participants from vulnerable groups. The author talks about the time and effort taken to gain access to the women, the role of the researcher as facilitator, challenges of interaction in the focus groups, and the relationship between research and therapy.

Reed, J. & Payton, R.K. (1997) 'Focus groups: Issues of analysis and interpretation', *Journal of Advanced Nursing*, 26, 765–71.

A valuable overview of key differences in the approaches to analysing qualitative data from focus groups and interviews.

References

Aldred, H., Gott, M. & Gariballa, S. (2005) 'A qualitative study to explore the impact of advanced heart failure on the lives of older patients and their informal carers', *Journal of Advanced Nursing*, 49 (2) 116–24.

Almack, K., Seymour, J. & Bellamy, G. (2010) 'Exploring the impact of sexual orientation on experiences and concerns about end of life care and on bereavement for lesbian, gay and bisexual elders', *Sociology*, 44 (5) 908–24.

Barnes, M. (2005) 'The same old process? Older people, participation and deliberation', *Ageing and Society*, 25, 245–59.

Brondani, M.A., MacEntee, M.I., Bryant, S.R. & O'Neill, B. (2008) 'Using written vignettes in focus groups among older adults to discuss oral health as a sensitive topic', *Qualitative Health Research*, 18, 1145–53.

Clark, D. & Seymour, J.E. (1999) *Reflections on Palliative Care: Sociological and Policy Perspectives* (Buckingham: Open University Press).

Clarke, A. & Seymour, J.E. (2010) '"At the foot of a very long ladder": Discussing the end of life with older people and informal caregivers', *Journal of Pain and Symptom Management*, 40 (6) 857–69.

Crockett, S.J., Heller, K.E., Merkel, J.M. & Peterson, J.M. (1990) 'Assessing beliefs of older rural Americans about nutritional education: Use of the focus group approach', *Journal of the American Dietetic Association*, 90, 563–7.

Davies, E. & Higginson, I. (eds) (2004) *Better Palliative Care for Older People* (Copenhagen: World Health Organization Europe).

De Laine, M. (2000) *Field-work, Participation and Practice: Ethics and Dilemmas in Qualitative Research* (London: Sage).

DoH (Department of Health) (2008) *End of Life Care Strategy – Promoting High Quality Care for All Adults at the End of Life* (London: Department of Health).

Duncan, M.T. & Morgan, D.L. (1994) 'Sharing the caring: Family caregivers' views of their relationships with nursing home staff', *The Gerontologist*, 34, 235–44.

Finch, J. (1987) 'The vignette technique in survey research', *Sociology*, 21, 105–14.

Gott, M., Seymour, J.E., Bellamy, G., Clark, D. & Ahmedzai, S.H. (2004) 'Older people's views about home as a place of care at the end of life', *Palliative Medicine*, 18, 460–7.

Hall, S., Longhurst, S. & Higginson, I.J. (2009) 'Challenges to conducting research with older people living in nursing homes', *BMC Geriatrics*, 9, 38. doi: 10.1186/1471-2318-9-38.

Hughes, A.O., Fenton, S. & Hine, C.E. (1995) 'Strategies for sampling black and ethnic minority populations', *Journal of Public Health Medicine*, 17, 187–92.

Keller, K.L., Sliepcevich, E.M., Vitello, E.M., Lacey, E.P. & Wright, W.R. (1987) 'Assessing beliefs about the needs of senior citizens using the focus group interview: A qualitative approach', *Health Education*, 18, 44–9.

Kendall, M., Harris, F. & Boyd, K. (2007) 'Key challenges and ways forward in researching the "good death": Qualitative in-depth interview and focus group study', *BMJ*, 334, 521–7.

Kitzinger, J. (1995) 'Qualitative research: Introducing focus groups', *British Medical Journal*, 311, 299–302.

Madriz, E. (2000) 'Focus groups in feminist research' in N.K. Denzin & Y.S. Lincoln (eds) *Handbook of Qualitative Research*, pp. 835–50 (Thousand Oaks: Sage).

Morgan, D.L. (1997) *Focus Groups as Qualitative Research* (London: Sage).

Morrow, E. (1997) 'Attitudes of women from vulnerable populations to physician and assisted death', *Journal of Clinical Ethics*, 8, 279–89.

Owen, S. (2001) 'The practical, methodological and ethical dilemmas of conducting focus groups with vulnerable clients', *Journal of Advanced Nursing*, 36, 652–8.

Pleschberger, S. (2007) 'Dignity and the challenge of dying in nursing homes: The residents' view', *Age and Ageing*, 36 (2) 197–202.

Reed, J. & Payton, R.K. (1997) 'Focus groups: Issues of analysis and interpretation', *Journal of Advanced Nursing*, 26, 765–71.

Sanders, C.M., Seymour, J.E., Clarke, A., Gott, M. & Welton, M. (2006) 'Development of a peer education programme for advance end-of-life care planning: An action research project with older adults', *International Journal of Palliative Nursing*, 12, 216–23.

Seymour, J.E., Almack, K., Bellamy, G., Clarke, A., Crosbie, B., Froggatt, K., Gott, M., Kennedy, S., Sanders, C. & Welton, M. (2009) *A Peer Education Programme for End of Life Care Education Among Older People and Their Carers* (Final report submitted to the Burdett Trust for Nursing). (Available from the author on request).

Seymour, J.E., Payne, S., Chapman, A. & Holloway, M. (2007) 'Hospice or home? Expectations about end of life care among older white and Chinese people living in the UK', *Sociology of Health and Illness*, 29 (6) 872–90.

Seymour, J.E., Sanders, C., Clarke, A., Welton, M. & Gott, M. (2006) *Planning for Choice in End-of-Life: An Educational Guide* (London: Help the Aged).

Seymour, J.E., Witherspoon, R., Gott, M., Ross, H. & Payne, S. (2005) *End of Life Care: Promoting Comfort, Choice and Well Being Among Older People Facing Death* (Bristol: Policy Press).

Seymour, J.E., Gott, M., Bellamy, G., Clark, D. & Ahmedzai, S. (2004) 'Planning for the end of life: The views of older people about advance statements', *Social Science and Medicine*, 59, 57–68.

Steinhauser, K.E., Clipp, E.C., McNeilly, M. Christakis, N.A., McIntyre, L.M. & Tulsky, J.A. (2000) 'In search of a good death: Observations of patients, families and providers', *Annals of Internal Medicine*, 132, 825–32.

Thornton, T. (1996) 'A focus group enquiry into the perceptions of primary health care teams and the provision of health care for adults with a learning disability living in the community', *Journal of Advanced Nursing*, 23, 1168–76.

Warren, L., Cook, J. & Clarke, N. (2003) 'Working with older women in research: Some methods based issues', *Quality in Ageing-Policy, Practice and Research*, 4 (4) 24–9.

WHO (World Health Organisation) (Europe) (2011) *Palliative Care for Older People: Better Practices* (Copenhagen: World Health Organisation).

Williams, R. (1990) *A Protestant Legacy: Attitudes to Death and Dying Among Older Aberdonians* (Oxford: Clarendon Press).

8
Using Online Methods to Interview Older Adults about Their Romantic and Sexual Relationships

Sue Malta

Introduction

Structural ageing of populations has been accompanied by a move towards a greater understanding of the long-term health care and housing needs of older adults. Unfortunately, however, their status and value to society remains in question. In Western cultures, at least, old age is typically viewed in a negative light (Birren & Schaie, 2006, p. 389), and ageing is seen as a process of 'inevitable decline and deterioration' (Friedan, 1993, p. 9), or as a time of 'decline, retreat, and withdrawal' (Dychtwald, 2005, p. 17). Ageism refers to negative generalisations and perceptions about older adults and is a fact of life for many older people (Age Concern, 2006; Gething et al., 2003). As a group, older adults have become used to being typecast as frail, confused and a 'burden' on society (de Vaus et al., 2003, p. 19; Hoyer, 1997, p. 39) thereby leading to research that further problematises the lives of older people and focuses on ways of 'dealing' with the burden of a growing ageing population.

Such negative generalisations and perceptions extend to all aspects of older adult lives, including sexuality. Consequently, late-life romance, both on- and offline, is a much neglected area of research (see Gott & Hinchliff, 2003 for discussion). Perhaps this is due to such ageist views and stereotypes which continue to portray older adults as asexual and, despite evidence to the contrary (Philbeck, 1997), as technologically incompetent. There is also a misperception that older adults are unwilling or reluctant to talk about intimacy and sexuality (Gledhill et al., 2008; Minichiello et al., 1996).

Historically, social research into relationships, including romance, love and sexuality has largely omitted older adults. For instance, the Australian Study of Health Relationships specifically excluded respondents over the age of 60 (Minichiello et al., 2003; Richters & Rissel, 2005); as did similar large-scale studies in the United Kingdom (UK) (Johnson et al., 2001) and the United States of America (USA) (Laumann et al., 1999). The reason for such generalised exclusion is hard to fathom. Perhaps older adults are perceived by 'researchers and ethics committees' as regarding the topic of sexuality as 'taboo and offensive' and therefore they are expected to be 'unwilling or unable' to talk about their sexuality and sexual needs, or indeed that 'they have no sexual needs' (Minichiello et al., 2003, p. 186). The presence of such stereotyping has extremely important negative repercussions, as it means that older adult needs and priorities are often omitted from vital national health, research and policy agendas (Gott & Hinchliff, 2003).

Research which has specifically looked at heterosexual older adults has tended to focus on their sexual functioning rather than the meanings they give to their loving relationships (see Gott & Hinchliff, 2003). Gott suggests that this is the case because many researchers attempt to treat sexuality as something that exists as an observable fact or a 'concrete phenomenon' which can be measured, whereas, in reality, its actual *meaning* is rarely ever explicated (Gott, 2005, p. 11, emphasis added).

Much of the research in this area has been quantitative and has centred mainly on medical issues such as declines in sexual activity over time, although there have been some recent exceptions (e.g. Waite et al., 2009). Connidis has argued that studies such as these do little to show how important sex actually is in the lives of older adults (2010, p. 61). As Weg noted almost 30 years ago, it is clear that:

> (…) the presence or absence of orgasm, masturbation, and intercourse, as the most reportable facts among the old as among the young, has overshadowed the emotional, sensual, and relationship qualities that give meaning… What has been largely missed are studies of changing love patterns with time (Weg, 1983, p. 8).

For these reasons research into older adult romance and sexuality is still in its infancy and very little is known about the ways in which they meet each other, commence dating, and consolidate their relationships into ongoing, intimate partnerships.

Dating and courtship in late-life relationships

In the late 1980s to the early 1990s, Bulcroft and colleagues conducted a series of studies highlighting the nature and functions of dating in later life, using both qualitative (interviews) and quantitative (survey) methods. They found there were important differences between older and younger daters, suggesting that while younger people date, for what the authors refer to as 'mate selection', older people date because they want to establish 'serious' (Bulcroft & Bulcroft, 1991, p. 246) and long-term romantic relationships, to alleviate loneliness and for the purposes of sexual fulfilment (ibid.). The earlier study by these researchers found that whilst relationships developed quickly and rapidly became sexual (Bulcroft & O'Connor, 1986a), very few couples provided ongoing instrumental support to each other in the way of housekeeping, health care and finances (Bulcroft & O'Connor, 1986b). The authors also found that very few of these dating relationships progressed to marriage (Bulcroft & Bulcroft, 1991; Bulcroft & O'Connor, 1986a), due to a need to remain independent and – especially for the women – a desire to avoid the caregiving role (Bulcroft & O'Connor, 1986a). In essence, these late-life couples maintained close, romantic and intimate relationships whilst living separately; a phenomenon which has since become known as living-apart-together (Levin & Trost, 1999).

A more recent qualitative study looked at dating amongst heterosexual older women in the US (Dickson et al., 2005). Providing additional support for the research by Bulcroft and colleagues, this research showed that although these women wished to sustain committed, intimate relationships, they did not want to cohabit or marry their partners (p. 78); again due to a strong desire to remain independent, both emotionally and financially, and a fear of being thrust into a nurse/caretaker role (p. 74). The authors suggested this created an 'atypical' dynamic in these later-life relationships in comparison to relationships in younger age groups. They argued that as younger daters were ultimately looking for life partners, they would feel the need to move their relationships towards more fixed, permanent entities (Dickson et al., 2005, p. 78); whereas this was not the case for older daters. The authors concluded that: 'marriage and commitment are prominent patterns in intimate relationships that later-life women seem to reject' (Dickson et al., 2005, p. 78). Although this study also touched on the sexual nature of some of the late-life relationships, the discussion was brief and did not provide enough evidence to make any conclusions about romance and sexuality in this cohort. Overall the limited research into dating and courtship in late life suggests there is a strong need/desire amongst older adults – and most especially women – to retain their independence and maintain separate lives from their romantic and intimate partners.

Older adults and computers

Whilst recent data indicates that older adults globally use the Internet less than their younger counterparts (Center for the Digital Future, 2009); other evidence demonstrates that older adults are making the digital conversion in increasing numbers (Ewing et al., 2008; Fox, 2004). The Australian Bureau of Statistics (ABS) reported that, whilst most age groups were experiencing small declines or a plateau in computer usage, the proportion of adults aged 50–64 years and 65 years and over using computers was continuing to grow; rising from 55 per cent and 21 per cent respectively in 2002 (ABS, 2005) to 66 per cent and almost 30 per cent respectively in 2007 (Ewing et al., 2008). Comparable results were found in the UK, where the number of retirees using the Internet has remained fairly stable at approximately 30 per cent since 2005 (Dutton & Helsper, 2007). In the US between 2000 and 2004, the percentage of seniors who accessed the Internet jumped to 22 per cent (Fox, 2004); with the figure increasing to 42 per cent by 2008 (Center for the Digital Future, 2009). This rate of increase is striking when one considers that only 2 per cent of the American elderly were online in 1996 (Pew Research Center for the People and the Press, 1996). Other countries have also exhibited this trend, with older adults (65 plus years) in Canada, New Zealand and Sweden echoing the American figures (45 per cent, 39 per cent and 38 per cent respectively; Center for the Digital Future, 2009).

But how do older adults use computers and the Internet? Older adults in the UK, USA and Australia use the Internet to send or read email, with many accessing health care or medical information (Fox, 2004), shopping, banking, paying bills and engaging in chat groups (Kiel, 2005), checking their genealogy, keeping up with the stock market and expanding their social networks (Adams et al., 2003), viewing or printing maps, checking the weather and viewing/posting photos and using/visiting the social networking site, Facebook (NielsenWire, 2009), checking online accounts, reading news and current affairs, as well as accessing travel, health and medical information (Australian Communications and Media Authority (ACMA), 2009), playing games, organising and storing photos, downloading music (Goodman et al., 2003) and, generally, searching for information (UK Ofcom, 2006). According to a report by Fox, once they make the transition to online technology, '...seniors are just as enthusiastic as younger users' (Fox, 2004, p. 3).

These studies provide clear evidence that older adults use the Internet in increasing numbers and perform online tasks similar to those of younger populations. Additionally, there is no evidence to suggest that older adults cannot engage in online interviews in much the same manner as other age groups who use the Internet. It is clear that online interviewing techniques are steadily increasing in acceptance as valid methods of conducting qualitative research (see for instance, Fielding et al., 2008; Hine, 2005). However, with very limited exceptions (see for example Xie, 2005), researchers in ageing have generally not considered online interviewing techniques – such as instant messaging or email – as suitable or even effective means for interviewing older adults. This is surprising, given the mounting evidence which shows that not only are older adults technologically competent (Fox, 2004), but that they also represent the fastest growing segment of Internet users (ABS, 2005; Ewing et al., 2008; Center for the Digital Future, 2009).

Researching older people's online dating

When considering social research into online romantic relationships, it is clear that they are rapidly becoming the 'norm' for most segments of the population – although the extent of older adults' involvement has still to be established. Whilst a plethora of anecdotal evidence (predominantly newspaper accounts) suggests that older adults are involved in such relationships, academic evidence is in relatively short supply. A Canadian study found 1.6 per cent of online daters were aged 60 plus (Brym & Lenton, 2001, p. 14), but most studies used younger sample populations, making it difficult to generalise to older adults (see for instance Donn & Sherman, 2002; Parks & Floyd, 1996; Underwood & Findlay, 2004; Whitty & Gavin, 2001). More recently, statistics supplied by *RSVP.com.au*®, an Australian online dating site, indicate that older adults (aged between 56 and 120 years) comprise 11 per cent of their membership database (Fairfax Digital Media, 2010), up from 8 per cent in 2008 (Fairfax Digital Media, 2008).

In terms of conducting research online, previous studies have shown that the online environment can be of particular benefit when sampling (and interviewing) hidden, vulnerable or marginalised populations, such as gay, lesbian, bisexual and transgender (GLBT) persons (Matthews & Cramer, 2008) and those with disabilities (Bowker & Tuffin, 2002, 2004). Engaging online allows disabled people, for instance, to overcome their everyday communication and mobility impairments, enabling them to interact 'outside the realm of [their] disability' (p. 335), which also allows them to be judged on the basis of their 'merits' rather than their disabilities (p. 330) – and thus to construct themselves as 'normal' online (p. 337). Furthermore, using the Internet to connect with (and to collect data from) socially marginalised respondents, allows respondents to remain at home in a safe, anonymous and non-threatening environment. Bowker and Tuffin have argued this not only furthers their ability to interact, but may well contribute to enhanced respondent disclosure and, as a consequence, the 'richness of the data gathered' (2004, p. 230).

Clearly not all older people are vulnerable or would see themselves as such and, indeed, those who have an online presence are far less likely to be so. Nonetheless, using the Internet to interview older adults is a realistic option, especially given their increasing online access and use of the technology. Additionally, discussing sensitive topics such as sex and sexuality via the Internet seems highly appropriate for this population group for all of the above reasons and, in particular, for maintaining anonymity. As already noted, there is very little qualitative research concerning older adults and online dating or offline dating for that matter. And with the exclusion of the study by Xie (2005) there is no research regarding how to conduct online interviews with older adults.

In 2006, the first of the so-called 80 million baby boomers in the USA turned 60 (Lawson, 2003, p. 26). Given that older adult Internet involvement is steadily increasing, it is interesting to speculate that there will be a large number of older adults wired up and looking for romance online:

> For decades, as a group, [the Baby Boomers] have challenged and pushed the limits of...norms; redefining the concepts of family and sexual expression (among many others)...we should anticipate increased changes and challenges in other social norms as this huge cohort moves into later life... Their technological competence and relative affluence will increase the potential of the Internet as a source of personal and social expression of sexuality in later life. Will we keep pace? (Adams et al., 2003, p. 413)

Accordingly, this chapter discusses online interviews with older adults, by showcasing a research project which investigated new late-life romantic relationships, online and offline. The online interviewing techniques used in the study, together with practical pointers on how difficulties were circumvented, are discussed in terms of recruitment strategies and interviewing (using instant messaging or email). The chapter is a reflection of my own research experience. Each section takes a step-by-step approach to describe how the participants were recruited, the online interviews conducted, and the data analysed. In addition, I draw on relevant research literature to cover issues pertinent to this line of research that were not always experienced in the course of the project.

The research project

My study explored the development of new late-life romances in a sample of older adults, aged 60 years of age and over residing in Australia: beginning with how and where these older adults met their new partners, their early dating experiences and the progression and consolidation of their new relationships into partnerships. I was particularly interested in the differences/similarities between relationships which began online and those that began face-to-face. My study further explored the notions of love, sex and intimacy – and whether such things actually existed in these new late-life relationships, as there was some ambivalence regarding these concepts in the literature about what they actually meant. Additionally my study also looked at whether these older adults' engaged in offline and online infidelity and cyber-sex.

   I recruited participants through a variety of means: by an online call for participants (CFP) which was placed on certain dating websites, by gaining publicity in local and interstate news media and radio, as well as in some senior-specific publications (*PrimeTime Seniors*® and *Fifty Plus*®) and through word-of-mouth referrals from personal networks, or through friends of other participants; in effect, creating a snowball sample.

The sample

My sample consisted of 45 heterosexual, community-dwelling, older adults aged 60 plus (range 60–92 years), consisting of 24 females and 21 males, who were currently engaged in or had recently been involved in a late-life romantic relationship, online or offline. Due to gender differences in population ageing (ABS, 2008), there are usually far more female participants in studies utilising older adult populations. In the current study this was not so, as there was only a slight difference in gender breakdown (see Table 8.1). This may be a reflection of the larger number of males versus females online generally, although this gap is steadily narrowing (e.g. Zamaria & Fletcher, 2008), or it may just be an artifact of the study itself. However further Internet-based studies are needed to assess this propensity.

   Two groups of older adults were interviewed: the Face-to-Face (F2F) Romance group (13 participants), who met in person through social

Table 8.1 Sample Characteristics and Internet Usage Statistics for the Online Romance versus the F2F Romance Group

	Total sample (n = 45)	
Description	Online Romance Group (n = 32)	F2F Romance Group (n = 13)
Gender (Females : Males)	16 : 16	8 : 5
Age Range (Years) Mean (Median)	60–76 65.5 (65)	63–92 71.5 (69)
Years Online Range Mean (Median)	1–20 10.5 (10)	8–17* 10.6 (10)*
Hours Online/Day Range Mean (Median)	0.2–10 3.6 (3)	0.43–4* 1.6 (1.25)*

*(n = 8)

groups/events and through family and friends, and the Online Romance group (32 participants), who met through online means and via dating websites in particular. The smaller sample size in the F2F Romance group was due to difficulties in recruiting offline. A breakdown of pertinent demographic details and Internet usage statistics by group has been included in Table 8.1.

Most of the older adults in this study were highly educated (from associated (two year) degree to postgraduate qualifications); were employed or had previously been employed in what could be described as typically middle-class occupations (academic, social worker, nurse, teacher, writer, for instance). Forty of the 45 participants were computer and Internet users. Countering the stereotype that older adults are technophobic, the number of years online plus the Internet usage statistics also indicated a high degree of familiarity and trust with the technology even for the group whose new late-life relationships began face-to-face. The findings also fit with previous research which suggests that early adopters of technology are more likely to be in higher status occupations (Dickerson & Gentry, 1983, p. 226) and wealthy (Fox et al., 2001, p. 3), and that those engaging in online romantic relationships are generally well educated (Wysocki, 1998, p. 435). There were, however, a range of experiences in terms of exposure to these technologies: five participants had never used a computer or the Internet before, and only became familiar with them as a means to pursue relationships online, whilst others had been using computers and the Internet for many years in both their working and personal lives.

Overview of interview methods

My study utilised four different online and offline qualitative interviewing methods: (i) face-to-face, (ii) telephone, (iii) synchronous computer-mediated communication (variously known as instant messaging, IM or private 'chat'[1]) and (iv) asynchronous electronic mail (email) correspondence. The remainder of the chapter concerns itself with the online recruitment methods and the online interviews only (for a more detailed discussion of face-to-face and telephone interviews see Malta, 2008).

I used the computer operating system Windows XP and conducted the IM interviews using standard, open-source (free), proprietary software, such as *Windows Live Messenger®*, *Yahoo®* or *Hotmail®* Instant Messaging programs. Of the older adults who chose this method to be interviewed, a relatively small number were unfamiliar with the technology and asked for some guidance in setting it up. I sent them an email with a simple, one-page list of instructions (accessed from the Internet), and which showed how to download the necessary software.[2] In most cases this proved unproblematic. These participants were pleased to have learnt a new skill and commented that they would now be able to IM online with their grandchildren. This outcome provides an excellent example of reciprocity arising from the research relationship (Harrison et al., 2001).

Recruitment strategies for sampling older adults online

As this was a new area of research for me, I decided to begin conservatively. I conducted an initial pilot study with *SeniorNet®*, one of the largest seniors' websites and organisations on the Internet (http://www.seniornet.org). Although it is located in and frequented predominantly by Americans, membership to this organisation is open to people all over the world. I chose this site because of its reputation as a senior-specific site and also because of the administrators' willingness to accommodate requests for research projects, as evidenced by the number of requests posted on *SeniorNet®*.

I sent off a polite request to the moderators/site administrators of *SeniorNet®* seeking permission to publish a study notice. This was granted after I supplied some detailed information as to the intention and scope of the research project, verification of my affiliation with a bona fide University, and upon my registration as a member with the website. I was then able to request the exact placement of the online notice. In this particular case it was included in the *SeniorNet®* Forum, under specific discussion threads such as 'Lifestyles', 'Relationships' and so on. Respondents to this online notice then replied directly to me, via my University email address, without any further need to go through the moderator. I then conducted semi-structured qualitative interviews via IM with this initial (pilot) sample, which consisted of five older American adults (aged 61–85 years). This data was not included in the final study.

The aim of this pilot project was to gain experience in conducting online interviews but it also provided the means to refine my interview questions. The interview schedule initially contained a series of questions about non-romantic friendship which the pilot interviews revealed to be unnecessary. These questions not only made the interview too lengthy but also confused participants as to the focus of the research. The questions were therefore omitted from the final schedule.

For the next stage of the project, I chose to focus my study within an Australian context and so I approached *RSVP.com.au®*, Australia's largest dating website and a senior-specific dating website *foreveryoungclub.com. au®* – although *Match.com®* or other similar sites such as *50YearsPlus.com®*, *Primesingles.net®* would have been just as suitable. Of all the recruitment techniques I used, online websites were the most successful source of older adult participants. This was particularly so for *RSVP.com.au®*, where registered users aged 60 years or more were targeted specifically by the site owners (Fairfax Digital) who sent out my CFP directly to their members. There were many more seniors-focused websites that, in hindsight, I could have used, rather than relying on participants from only one or two sites. One online study, for instance, which interviewed health in-formation seekers of all ages, placed advertisements on 20 different web-sites related to the study topic (healthy eating, fitness and general health; Kivits, 2005). However, as I ended up with more than enough respondents, casting the net too wide might have made my study unwieldy.

Interested respondents contacted me via my University email account or my mobile phone. I received more requests (76 in total) than anti-cipated (I expected about 40–50). However, many participants dropped out of their own accord (they either stopped communicating (19) or decided to withdraw (four) from the study) or had to be excluded because they did not meet the study criteria (eight) (they were less than 60 years old or they were not currently or recently involved in a new late-life relation-ship, in other words, they were still looking). The final sample comprised of 45 people, 40 of whom were computer literate. These factors alone served to provide additional justification for the study in that they confirmed: (1) that many older adults had access to and used the Internet on a regular basis and (2) that many older adults were involved in or interested in late-life romance – and used the Internet to access dating

Table 8.2   Interview Method by Relationship Group (n)

Interview Method	Online Romance Group	F2F Romance Group	Totals N
Face-to-Face	1	9	10
Telephone	5	–	5
Online: IM	23	3	26
Online: Email	3	1	4
Totals	**32**	**13**	**45**

websites, and (3) a willingness to take part in research involving online methods.

Offering a range of interview choices

Allowing participants to decide the interview medium they prefer has been purported to increase both retention rates and rapport between the researcher and researched (Kazmer & Xie, 2008, p. 273). There was no reason to believe that older adults would behave any differently, and accordingly they were given a choice of four different semi-structured qualitative interview methods: (i) face-to-face, (ii) telephone, and online via (iii) IM and (iv) email correspondence. As a result, the interview mode which each one chose did not always reflect which relationship group they belonged to. Table 8.2 shows a breakdown of the interview modes and how they corresponded to the two different relationship groups. Overall, 30 older adults (from a total of 45) chose to be interviewed online, either by IM or email, again, indicating a high level of familiarity and trust in computers and technology. The four participants who chose to be interviewed by email did so because they were either ill or lacked the time (due to work constraints) to commit to an interactive interview, and preferred the flexibility that email afforded them, that is, they could answer in their own time.

Interviews lasted between one and two hours, with the exception of the four email interviews. Surprisingly there was very little time difference in the length of the face-to-face, telephone and IM interviews, although some researchers have noted that online interviews can take longer than face-to-face ones due to the time it takes to type in questions (Kazmer & Xie, 2008).

Instant messaging (IM) interviews

Two striking reasons for using IM for interviews are geographical flexibility and the production of instant transcripts. In the first instance, researchers and respondents can be located anywhere in the world, subject only to time differences between countries and regions (Bowker & Tuffin, 2004; Opdenakker, 2006). This was certainly true for my project as I was able to conduct my pilot study interviews with participants who were located in the USA with much the same ease as those who were located within Australia, and with little expenditure, apart from the usual Internet connection charges. In the second instance, the generation of instant transcriptions means significant time and even more cost savings for the researcher (Bowker & Tuffin, 2004). As 30 of the 45 interviews in my study were conducted by IM or email – and subsequently created their own transcripts – this amounted to a considerable difference to project costings.

Other reasons for using IM for interviewing concern issues of participant comfort, familiarity and disclosure. For instance, Hammersley and Atkinson (1995, p. 150) argued that interviewing people within their own familiar surroundings is 'the best strategy' because it enables participants to be more comfortable and relaxed than they would be in unfamiliar locations. Although these researchers were talking more specifically about face-to-face interviewing, I found these comments held true for conducting IM interviews and they also fit with the online work of Bowker and Tuffin (2002, 2004) reported earlier. Couch and Liamputtong (2008) have specifically recommended interviewing online daters by IM, because it is in common usage amongst online daters and, therefore, it offers a 'mode of communication [which is both] relevant and appropriate' (p. 270). In an early example of an online study, Hamman (1997) suggested that the use of online interviews allows participants to be 'more candid' than they would be in face-to-face interviews when discussing such issues as sexuality (p. 6). All these factors were pertinent to my study. In addition to exploring an under-researched method of investigation, I was motivated by curiosity to see whether interviewing by IM was, in fact, a feasible method to employ with older adults.

There is one major problem associated with conducting online interviews, which concerns categorising the demographics of the sample, as it is possible that respondents lie about this information. This is an issue relevant to all methods, but particularly to non face-to-face ones. I found that, if alert, I could easily check for discrepancies whilst the interview was in progress, or later via follow-up emails. In countering this issue, Stieger and Goritz (2006) conducted a unique online study which used three different methods to cross-check information (a web-based survey, self-reports and IM interviews) and found that the possibility of receiving bogus data in IM interviews is actually relatively minor. In my project, there was only one such occasion where the responses I received did not seem to reflect what was ostensibly an older adult male perspective. I was surprised by his responses, which were written in a shorthand form akin to mobile phone text messaging. This in itself was not enough to indicate proof of an age disparity, but as he was the only one from whom I received such responses, it put me on the alert. This participant then proceeded to take a very active role in the interview in that he asked more questions than he answered, which increased my suspicion. Consequently, I asked him what year he was born. It took him some time to answer this question, although he claimed to be 65 years old. This confirmed my suspicions and I politely terminated the interview shortly thereafter.

Semantics, scheduling and set-up issues, including time constraints and software needs

When this research project was developed, sceptics argued that older adults would be difficult to reach online. And for the pilot project this initially proved true. However, it was soon clear that this was to do with the wording of the CFP, rather than any reluctance by older adults to be involved in an online interview project. The original CFP asked for older adults who were interested in talking about their 'intimate' relationships. Not one response was received. The notice was subsequently reviewed, the word 'intimate' was replaced with 'romantic' and the CFP was re-posted. A number of participants then came forward.

The minor semantic difference in the original CFP suggested that the use of the word 'intimate' had been taken to be synonymous with the word 'sex' which, in turn, appeared to create a hesitancy to participate amongst online older adults. A dilemma thus presented itself. Would this word-sensitivity in respondents ('intimate' versus 'romantic') affect the scope and quality of the data collected? For instance, would this hesitancy be reflected in the details that older adults would be willing to disclose about their relationships? Were older adults reluctant to talk about matters concerning sexuality, as some researchers have suggested (Gledhill et al., 2008) and would this reluctance thus render the project unfeasible? These fears proved to be unfounded. Once older adults agreed to participate they were more than comfortable to talk about *all* aspects of their romantic relationships – intimate or otherwise (see Chapter 2).

When conducting the IM interviews, I needed to take account of different time zones – as there can be as much as a three hour time difference between regions of Australia (e.g. Melbourne to Perth). This meant that for some locations there was potentially a small window of opportunity to schedule and conduct interviews. However, I found it relatively easy to negotiate an interview time which was acceptable to myself and my participants, as did Hinchliff & Gavin (2009), whilst Markham (1998) reported scheduling difficulties.

As the researcher, it was my ethical duty to ensure that participant identity was protected at all times and to maintain the confidentiality and privacy of the online interviews. I achieved this by ensuring that I conducted the IM interviews in a secure, private location, where it was not possible for the computer screen to be overlooked by anyone in the vicinity, or even the casual passer-by. In a study like this which concerned such delicate matters as love, sex and intimacy, these issues seemed to me to be even more paramount. It was important that I was able to reassure my respondents that they were able to 'speak' freely and safely, without being compromised.

The process of IM interviews was really very simple. All it took to set up was for both participants and myself to have access to the same instant messaging program, and then one or the other would send an 'Invitation to IM'. Once the connection was established, a dialogue box appeared. Usernames indicated who was online and text appeared next to each username, indicating that one or the other was typing a message (see Figure 8.1).

Voida et al. (2004, p. 1345) have said that the instantaneous nature of this form of communication (IM) facilitates 'rapid, fluid exchanges' which are constrained only by a person's typing and reading speed. I found this to be true, and further, that the IM interviews resembled face-to-face or phone interviews, in that they were one-on-one between me and the participant, they occurred in real-time, and they mimicked the normal processes of a face-to-face back-and-forth conversation.

The online interview relationship

I made initial arrangements regarding scheduling interviews via the telephone or email prior to establishing the IM connection (Kivits, 2005). This preliminary contact allowed my participants (and myself) time to air any questions/concerns. I found this early contact required careful treatment, as it functioned in the same manner as 'first impressions'

Figure 8.1   Representation of an IM dialogue in *Windows Live Messenger®*

8/03/2008	12:56:34	Maria	hi sue
8/03/2008	12:58:20	Maria	i'm ready, are you?
8/03/2008	12:58:47		Sue has been added to the conversation
8/03/2008	12:58:47	Sue	Yes, Maria, I am
8/03/2008	12:59:18	Sue	Did you get my email this morning with the disclosure?
8/03/2008	12:59:47	Maria	yes, is it sufficient that i give consent here?
8/03/2008	1:00:03	Sue	absolutely
8/03/2008	1:00:16	Maria	ok, then i consent to participating in the study
8/03/2008	1:00:21	Sue	Thank you Maria
8/03/2008	1:01:07	Sue	I will just start off with a couple of housekeeping points, if that's okay? I would just like to remind you that this interview is entirely confidential and that you will not be able to be identified in any manner. I am also going to be copying the text to use in the future. Is that alright with you Maria?
8/03/2008	1:01:20	Maria	yes
8/03/2008	1:01:28	Sue	My study is of older adults, therefore I need to keep track of participants' ages. Can you please tell me how old you are?
8/03/2008	1:01:40	Maria	67
8/03/2008	1:01:45	Maria	i'll be 68 in october
8/03/2008	1:02:08	Sue	Thank you for that Maria
8/03/2008	1:02:13	Sue	As gender is also an important part of my study, could you please confirm whether you are a male or a female?
8/03/2008	1:03:00	Maria	last time i looked i was a female – LOL!

given off in initial face-to-face contact, and facilitated the beginning of what was, in effect, an 'interpersonal' relationship between my research participants and me (Kivits, 2005). I also found that, with the absence of face-to-face cues, the written word could sometimes cause me to appear as a no-nonsense and abrupt personality, rather than an approachable interviewer. I think it is important that researchers using online methods be aware of this, as striking the right balance between professionalism and approachability under such circumstances, can sometimes be difficult. This initial contact should not be under-estimated, particularly when working with unfamiliar samples.

As in the study by Kivits (2005), the research relationship that was established between myself and the participants in this text-only environment, was generally warm and friendly although, as noted, in a minority of interviews it could be described as mostly professional. As in face-to-face and telephone interviews, rapport was established by using instances of self-disclosure and the use of humourous asides. Certainly when I disclosed my mature-aged status (I was 51 at the time of the interviews), this helped put participants at ease. As one older adult male who was being interviewed by IM and who was initially uncomfortable remarked: 'I don't think anyone of a much younger age would understand what I have to say'. However, many of the older adults 'talked' to me as if I was a peer well before this personal disclosure.

Online interviews have been postulated to be less affected by social desirability factors (Couch & Liamputtong, 2008; Tatano Beck, 2005), as well as an absence of interviewer/interviewee effects, such as the power dynamics inherent in traditional research methods, which may allow the researched to be on a more equal footing with the researcher (Fox et al., 2007; Meho, 2006). This seemed pertinent to my study, as I felt less as though I was conducting an interview, and more as though I was having a conversation, a feeling which developed more strongly as the interviews progressed. I would liken this to a feeling of equality developing between myself and the participants.

In terms of broaching such sensitive subjects as love, sex and intimacy, I found the IM interviewing mode was relatively stress-free and less awkward for both the participants and me, compared to the face-to-face and telephone interviewing modes. Perhaps this was due to participants feeling less self-conscious in the relative anonymity of the online environment which some researchers have said facilitates lower inhibition and can lead to greater disclosure (Bowker & Tuffin, 2004; Mesch & Beker, 2010) – although others have found the opposite (Davis et al., 2004). My study would appear to support this contention, as many of my older adult participants were extremely frank in their discussions of love, sex and intimacy. They also commented on the ease with which the interviews were conducted, and said they felt very comfortable being interviewed anonymously in the location of their choice (in most cases this was their own homes, although a small number were located at private business premises).

Although some researchers have noted the use of 'emoticons' such as smiley faces (☺) in online interviewing (Fontes & O'Mahony, 2008;

Opdenakker, 2006), their presence in my study was very limited. However, abbreviations such as 'LOL' (laugh out loud), 'OK' (okay) and 'tx u' (Thank you), for example, appeared to be fairly common. Whether this was because the participants were unfamiliar (or uncomfortable) with emoticon use or whether it was because older adults as a whole prefer text only as opposed to visual representations, is difficult to tell.

Once the IM interview was underway, I found that the essential requirement was to be, mostly, patient. As noted by Voida et al. (2004) some people type more slowly than others, some conduct more than one IM conversation at a time, and yet others perform other computer-related tasks simultaneously. Suler (1997) has referred to these delays as conversational 'hiccups'. At the time I conducted the interviews, I found such 'hiccups' were not difficult to pinpoint or deal with and everyone I interviewed online was familiar with them. Participants also sometimes said they were answering the phone or someone was at the door. Others would just go 'silent' for a while. When this happened, I would minimise the IM screen, turn up the computer speakers and wait for the audible beep which would tell me that my interviewee was back. I regarded these delays as beneficial, as they often acted as ideal mini-breaks in the interview conversation, allowing time for me to regroup and reflect.

Managing, copying and saving instant transcripts

The most difficult task I encountered at this stage related to the number of 'screens' or programs that needed to be open simultaneously, and my ability to multi-task between them. For instance, the computer 'desktop' displayed the email program and IM Box and the Word document contained the interview schedule and a new Word document into which to copy and paste the interview responses (see Figure 8.2). I also kept on hand a hard copy of the interview schedule, together with any notes from previous email or phone contact with the interviewee. This meant I was often moving backwards and forwards between many different online (and offline) documents at once, whilst trying to keep track of the interview by reading the transcript as it developed.

In terms of saving a permanent record of the data, I found it was necessary to perform a series of 'Selects' and 'Saves' after each question and answer. That is, I used 'Select All' and 'Copy' within the IM Box and then transferred the information to the open Word document, where I performed another 'Select All' and then 'Paste'. In this manner, the dialogue contained in the Word document was continually

Figure 8.2   Simulated representation of a computer desktop showing three open documents involved in an IM interview: the IM box, the Word document and the interview transcript

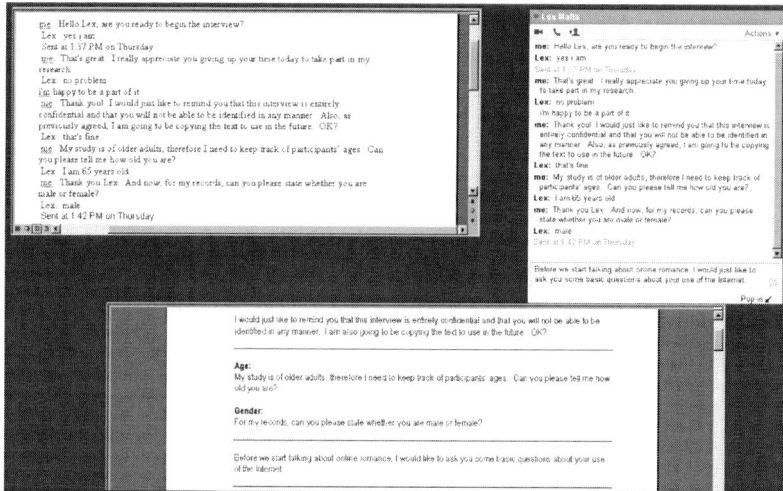

updated without any overlaps or doubling up. At the end of each 'Paste', I would then 'Save' the document to ensure none of the data was lost.[3]

It is necessary at this stage to highlight the importance of keeping detailed records. For me, this took the form of a chart, which indicated the names of contacts, their assigned pseudonyms, their location and relevant time zone, the scheduled time of interview and whether it was completed or not. With so many contacts and emails going back and forth, I sometimes found it difficult to remember who was who and when the interviews were scheduled. Having a chart to refer to, helped this process enormously. Keeping such a record (and the additional notes) helped me make a more nuanced and informed write up upon completion of the study (see Figure 8.3).

Email interviews

I found that conducting interviews by email affords many of the same advantages as using IM, not least of which is that it is easy to use, and produces a record of the interview. In contrast to IM, interviewing by email is achieved asynchronously, thus, it is not subjected to the vagaries of scheduling that can sometimes accompany IM and other methods (Meho, 2006).

Figure 8.3   Representation of a chart used to track the interviews in the study

Name	Psuedo	Age	F/M	Date	Time/ Locate/ Time Diff	Interview Length	Agreed to follow-up?	Interview Notes
XXXX	Maria	67	F	8 March 2008	2.30 pm Melb	2.30 pm–3.37 pm (67 mins)	Yes	Humourous
XXXX	Douglas	72	M	17 March 2008	8.30 am Melb	8.35 am–10.11 am (96 mins)	Yes	Not quite finished (slow typing skills) rescheduled after Easter 25 March 8.30 am
XXXX	Natalie	76	F	18 March 2008	7.00 pm Melb = 4.00 pm Perth	7.02 pm–8.43 pm (81 mins)	Yes	Copy of thesis

There are two methods of using email for interviewing: embedded and attached. The first involves embedding each interview question (or a small number of questions) within each email message. This results in a number of emails being exchanged over a period of time. Due to the time delay between sending emails and receiving replies, some researchers have noted unusually long time frames for their research projects. For instance, McCoyd and Schwaber Kerson (2006) reported that their email interviews typically required 8 to 14 interactions between themselves and their participants. Others have reported email interview periods that ranged from weeks or months (Bowker & Tuffin, 2004) to even as long as one year (Kivits, 2005). As I was constrained by time factors, I opted to use email attachments to conduct the interviews.

I also chose the attached email method because I preferred to receive a fully realised interview transcript, rather than to negotiate the back and forth between multiple emails. Furthermore, this approach also broadly mirrored the other interview methods I used in that the interviews were conducted in a discrete time frame, with participants staying focused on the topic. As Kazmer and Xie (2008) have noted, when sending questions embedded in many different emails, it is necessary to keep track of all the email responses, and then at some stage to compile them all into one document. Under these circumstances, it is always possible for data to be labelled inappropriately or for it to become mislaid. Consequently, after initial contact was established, I sent an email to my older adult participants with both the informed consent and the full interview schedule attached as Word documents.

Questions which are attached to emails have previously been shown to result in lower response rates than those which are embedded within emails (Dommeyer & Moriarty, 2000). Meho (2006) suggested this is because attachments present too many obstacles for study participants. For instance, people lose interest when they have to take extra steps to download and save a document especially, or if they do not possess a 'strong interest' in responding they may drop out (Meho, 2006, p. 1290). I did lose one participant after sending the email with its attachments and it may well have been for this reason, but as the respondent did not reply to follow-up emails it is difficult to say with any certainty.

I used the same interview schedule template which I used for the other three interview methods. Participants then typed their responses directly into the interview schedule at their own leisure and then emailed their responses back to me. I usually received the completed transcripts within a day or two, and the longest response time was only five days.

I asked all participants – both offline and online, and regardless of interview mode – if they were willing to be contacted if clarification was needed on any point. All agreed, so where responses in the email interviews were a bit thin, or if a question was missed, I was able to revisit certain points, check any ambiguities and seek elaboration where needed. In doing so, the points identified earlier by Kazmer and Xie (2008), regarding the organisational rigour required in collecting data from various online methods, came into their own.

Dynamics of IM versus email interviews

Answers received during the IM interviews tended to be in-depth and rich in qualifications, whereas primary responses to the email interviews were much shorter and succinct. This initial difference appeared to be due more to the interactive nature of IM communication, which mirrored a real-life telephone or face-to-face conversation, albeit in text, whereas with the asynchronous nature of attached email interviews the conversational ebb and flow was lost. Meho (2006) has argued, that this is not necessarily a negative outcome. He contends that email interviews actually afford participants the time to be more reflective of their replies, as well as providing them with the opportunity for editing of responses, making for more focused answers. As a result, although email interviews may be shorter than their IM counterparts, and may appear to contain less data, the data obtained can be just as rich and valuable as IM interviews. My study certainly reinforced this finding, as the level of disclosure in the email interviews was similar to that found in the IM mode. Additionally, in response to a question which stated 'Is there anything I haven't asked you that I should have?' three of the four participants who chose the email method included extra comments at the end of their interview schedules, whereas this occurred with only one of the IM interviews.

It was clear from my study that interviewing by email using attached questions could, potentially, be less rewarding an exercise for the researcher, as I felt it was sometimes difficult to establish the same connection/relationship as was evidenced in the face-to-face, phone and IM interviews. This point has been reinforced by the work of other researchers, who previously used the embedded email method and reported well-established interview relationships (Bowker & Tuffin, 2002, 2004; Kivits, 2005; McCoyd & Schwaber Kerson, 2006). However, as my older adults who chose this method appeared to find the experience rewarding, this is really a moot point.

Conclusion

The perception that older adults are technophobic has often precluded their inclusion in studies of Internet-based research. It is clear from the research that I present here that an easily sampled group of older adults were familiar with, and had trust in, the technology, as evidenced by their willingness to be interviewed via online methods.

Given this project was centred on investigating new late-life relationships initiated through Internet dating websites, online research methods provided the most appropriate means to explore the phenomenon. Furthermore, two-thirds of the older adults in the study chose to be interviewed by IM or email in preference to other more traditional modes of interviewing, and enjoyed the process. This adds weight to the argument that online interviewing is both a viable and practical option for conducting qualitative research with older adult populations, and that it also generates rich data.

This chapter has illustrated the many benefits of using the Internet and online interviewing for researching older adults. IM and email provide an effective means to access older adult populations who might normally be difficult to reach, they are easily achievable, are both time- and cost-effective, and also allow for the discussion of topics which at times may be embarrassing or confrontational in face-to-face and telephone interviews. Online interviews thus provide an important tool for researchers, particularly those studying potentially sensitive topics such as romance and sexuality.

The automatically generated transcripts enabled access to the data at a much earlier stage than with the face-to-face and telephone interviews. This early contact with the fully realised data is an additional and, until now, largely unacclaimed benefit of using online interviews. It allows the researcher to have immediate and continuing access to and engagement with the data, thereby facilitating immersion and consolidating knowledge at a much earlier stage than with methods relying on transcription.

Face-to-face interviews have traditionally been seen as 'the quintessence of qualitative research' (Seymour, 2001, p. 155), and face-to-face communication has long been regarded as the 'gold standard against which' online interaction is often judged (Hine, 2005, p. 4). Nevertheless, studies using online interviewing techniques are increasing rapidly, in line with the increasing numbers of people who have access to and use the Internet. As older adults have been shown to be the fastest growing population in this respect, it is time that researchers championed the use of such technology for their own research agendas, and by doing so help overcome stereotypes of older people.

Notes

1 To avoid confusion, instant messaging will hereafter be referred to as 'IM'.
2 It should be kept in mind that this exploratory study was conducted in 2007 and 2008. Since that time technology has changed considerably. It is now possible to

IM easily within most standard, publicly available email programs, such as *Yahoo*® or *Gmail*®, without having to bother with downloading additional software or without setting up a new account or username.

3   Current IM programs automatically save a copy of any exchange taking place.

Annotated reading list

Davis, M., Bolding, G., Hart, G., Sherr, L. & Elford, J. (2004) 'Reflecting on the experience of interviewing online: Perspectives from the Internet and HIV study in London', *AIDS Care*, 16 (8) 944–52.

This article discusses both the upside and pitfalls of using IM interviews. It also provides an example of an IM interview dialogue. Though the authors do not have an altogether positive view of the method, the article provides a good basis for researchers new to the technique.

Hinchliff, V. & Gavin, H. (2009) 'Social and virtual networks: Evaluating synchronous online interviewing using Instant Messenger', *The Qualitative Report*, 14 (2) 318–40.

This article explores the authors' experiences with using IM interviews for the first time. It is useful in that it provides evidence from a mixed-methods study albeit with a small sample of undergraduate students.

Kazmer, M.M. & Xie, B. (2008) 'Qualitative interviewing in Internet studies: Playing with the media, playing with the method', *Information, Communication & Society*, 11 (2) 257–78.

This article provides a detailed discussion comparing four different types of interview techniques, including email and IM. It provides one of the very few available examples of using these methods with older adult populations. Again, it is a good backgrounder for researchers new to these methods.

Kivits, J. (2005) 'Online interviewing and the research relationship' in C. Hine (ed.) *Virtual Methods: Issues in Social Research on the Internet* (Oxford: Berg Publishers).

Kivits details a research study using embedded email interviews. The chapter contains excerpts from the actual email exchanges between the researcher and participants, which provide interesting examples of how a research relationship can be established in the online environment. The examples also highlight the richness of the data gathered with the email method. Although now dated, the book itself is a valuable resource for researchers contemplating using online methods.

Stieger, S. & Goritz, A.J. (2006) 'Using Instant Messaging for Internet-based interviews', *Cyberpsychology & Behavior*, 9 (5) 552–9.

For researchers requiring reassurance on issues of validity when using IM interviews, this article details contact rates, response rates and retention rates. Steiger and Goritz compare self-reports, actual behaviour and external data to show that the possibility of receiving false data in IM interviews is relatively rare.

References

ABS (Australian Bureau of Statistics) (2005) 'Use of information technology by older people', Catalogue No: 1301.0. in *Year Book Australia, 2005* (Canberra: Commonwealth Government).

ABS (2008) *Population by Age and Sex, Australian States and Territories* Catalogue No: 3201.0 (Canberra: Commonwealth Government).

ACMA (Australian Communications and Media Authority) (2009) *Use of Digital Media and Communication by Older Australians.* http://www.acma.gov.au/WEB/STANDARD..PC/pc=PC_311711_date accessed 16 December 2009.

Adams, M.S., Oye, J. & Parker, T.S. (2003) 'Sexuality of older adults and the Internet: From sex education to cybersex', *Sexual and Relationship Therapy*, 18 (3) 405–15.

Age Concern (2006) *Synopsis: How Ageist is Britain? Ageism: A Benchmark of Public Attitudes in Britain* (London: Age Concern Research Services).

Birren, J.E. & Schaie, K.W. (2006) *Handbook of the Psychology of Aging* (Burlington: Elsevier Academic Press).

Bowker, N. & Tuffin, K. (2002) 'Disability discourses for online identities', *Disability & Society*, 17 (3) 327–44.

Bowker, N. & Tuffin, K. (2004) 'Using the online medium for discursive research about people with disabilities', *Social Science Computer Review*, 22 (2) 228–41.

Brym, R.J. & Lenton, R.L. (2001) *Love Online: A Report on Digital Dating in Canada* (Toronto, Canada, Funded by MSN.CA: 1–54).

Bulcroft, R.A. & Bulcroft, K. (1991) 'The nature and functions of dating in later life', *Research on Aging*, 13 (2) 244–60.

Bulcroft, K. & O'Connor, M. (1986a) 'The importance of dating relationships on quality of life for older persons', *Family Relations*, 35, 397–401.

Bulcroft, K. & O'Connor, M. (1986b) 'Never too late', *Psychology Today*, 20 (6) 66–9.

Center for the Digital Future (2009) *The 2009 Digital Future Project – Year Eight Report: World Internet Project (WIP)* (California: USC Annenberg School for Communication).

Connidis, I.A. (2010) *Family Ties & Aging* (Thousand Oaks: Pine Forge Press).

Couch, D. & Liamputtong, P. (2008) 'Online dating and mating: The use of the Internet to meet sexual partners', *Qualitative Health Research*, 18, 268–79.

Davis, M., Bolding, G., Hart, G., Sherr, L. & Elford, J. (2004) 'Reflecting on the experience of interviewing online: Perspectives from the Internet and HIV study in London', *AIDS Care*, 16 (8) 944–52.

de Vaus, D., Gray, M. & Stanton, D. (2003) *Measuring the Value of Unpaid Household, Caring and Voluntary Work of Older Australians* (Australian Institute of Family Studies, Research Paper No. 34 (October)).

Dickerson, M.D. & Gentry, J.W. (1983) 'Characteristics of adopters and non-adopters of home computers', *Journal of Consumer Research*, 10 (September) 225–35.

Dickson, F.C., Hughes, P.C. & Walker, K.L. (2005) 'An exploratory investigation into dating among later-life women', *Western Journal of Communication*, 69 (1) 67–82.

Dommeyer, C.J. & Moriarty, E. (2000) 'Comparing two forms of e-mail survey: Embedded vs. attached', *International Journal of Market Research*, 42 (1) 39–50.

Donn, J.E. & Sherman, R.C. (2002) 'Attitudes and practices regarding the formation of romantic relationships on the Internet', *Cyberpsychology & Behaviour*, 5 (2) 107–23.

Dutton, W.H. & Helsper, E.J. (2007) *The Internet in Britain* (Oxford: Oxford Internet Surveys, University of Oxford).

Dychtwald, K. (2005) 'Ageless aging: The next era of retirement', *The Futurist*, 39 (4) 16–21.

Ewing, S., Thomas, J. & Schiessl, J. (2008) *CCi Digital Futures Report: The Internet in Australia* (Hawthorn: ARC Centre of Excellence for Creative Industries and Innovation, Institute of Social Research, Swinburne University of Technology).

Fairfax Digital Media (2008, 2010) *RSVP.com.au. Where More Australians Meet* (Personal Communication).

Fielding, N.G., Lee, R.M. & Blank, G. (eds) (2008) *The Sage Handbook of Online Research Methods* (London: Sage Publishers).

Fontes, T.O. & O'Mahony, M. (2008) 'In-depth interviewing by Instant Messaging', *Social Research Update*, 53, 1–4.

Fox, F.E., Morris, M. & Rumsey, N. (2007) 'Doing synchronous online focus groups with young people: Methodological reflections', *Qualitative Health Research*, 17 (4) 539–47.

Fox, S. (2004) *Older Americans and the Internet* (Washington DC: PEW Internet & American Life Project Report).

Fox, S., Rainie, L., Larsen, E., Horrigan, J., Lenhart, A., Spooner, T. & Carter, C. (2001) *Wired Seniors* (Washington D.C.: PEW Internet & American Life Project Report).

Friedan, B. (1993) *The Fountain of Age* (London: Jonathan Cape).

Gething, L., Gridley, H., Browning, C., Helmes, E., Luszcz, M., Turner, J., Ward, L. & Wells, Y. (2003) 'The role of psychologists in fostering the wellbeing of older Australians', *Australian Psychologist*, 38 (1) 1–10.

Gledhill, S., Abbey, J. & Schweitzer, R. (2008) 'Sampling methods: Methodological issues involved in the recruitment of older people into a study of sexuality', *Australian Journal of Advanced Nursing*, 26 (1) 84–94.

Goodman, J., Syme, A. & Eisma, R. (2003) 'Older adults use of computers: A survey' in *Proceedings Human-Computer Interaction* (Bath, Sept 2003). http://www.dcs.gla.ac.uk/~stephen/ research/utopia/papers /2003_bcs_hci/paper.pdf

Gott, M. (2005) *Sexuality, Sexual Health & Ageing* (Berkshire: Open University Press).

Gott, M. & Hinchliff, S. (2003) 'How important is sex in later life? The views of older people', *Social Science & Medicine*, 56, 1617–28.

Hamman, R. (1997) 'The application of ethnographic methodology in the study of cybersex', *Cybersociology*, 1 (October 10) http://www.cybersociology.com/files/1_1_hamman.html, date accessed 7 August 2008.

Hammersley, M. & Atkinson, P. (1995) *Ethnography: Principles and Practice*, 2nd edn (London: Routledge).

Harrison, J., MacGibbon, L. & Morton, M. (2001) 'Regimes of trustworthiness in qualitative research: The rigors of reciprocity', *Qualitative Inquiry*, 7 (3) 323–45.

Hinchliff, V. & Gavin, H. (2009) 'Social and virtual networks: Evaluating synchronous online interviewing using Instant Messenger', *The Qualitative Report*, 14 (2) 318–40.

Hine, C. (2005) *Virtual Methods: Issues in Social Research on the Internet* (Oxford: Berg Publishers).

Hoyer, W.J. (1997) 'Positive contributions of the elderly to society: A multi-disciplinary perspective', *Supplement to the Australasian Journal on Ageing*, 17, 39–41.

Johnson, A.M., Mercer, C.H., Erens, B., Copas, A.J., McManus, S., Wellings, K., Fenton, K.A., Korovessis, C., Macdowall, W., Nanchahal, K., Purdon, S. & Field, J. (2001) 'Sexual behaviour in Britain: Partnerships, practices and HIV risk behaviours', *The Lancet*, 358 (9296) 1835–42.

Kazmer, M.M. & Xie, B. (2008) 'Qualitative interviewing in Internet studies: Playing with the media, playing with the method', *Information, Communication & Society*, 11 (2) 257–78.

Kiel, J.M. (2005) 'The digital divide: Internet and e-mail use by the elderly', *Medical Informatics and the Internet in Medicine*, 30 (1) 19–23.

Kivits, J. (2005) 'Online interviewing and the research relationship' in C. Hine (ed.) *Virtual Methods: Issues in Social Research on the Internet* (Oxford: Berg Publishers).

Laumann, E.O., Paik, A.M.A. & Rosen, R.C. (1999) 'Sexual dysfunction in the United States: Prevalence and predictors', *Journal of the American Medical Association*, 281 (6) 537–44.

Lawson, W. (2003) 'Aging's changing face', *Psychology Today* (July/August), 26.

Levin, I. & Trost , J. (1999) 'Living apart together', *Community, Work & Family*, 2, 279–94.

Malta, S. (2008) 'Intimacy and older adults: A comparison between online and offline romantic relationships' in T. Majoribanks et al. (eds) *Reimagining Sociology* (University of Melbourne: Annual Conference of the Australian Sociological Association).

Markham, A.N. (1998) *Life Online: Researching Real Experiences in Virtual Space* (Walnut Creek: Altamira Press).

Matthews, J. & Cramer, E.P. (2008) 'Using technology to enhance qualitative research with hidden populations', *The Qualitative Report*, 13 (2) 301–15.

McCoyd, J.L.M. & Schwaber Kerson, T. (2006) 'Conducting intensive interviews using email: A serendipitous comparative opportunity', *Qualitative Social Work*, 5 (3) 389–406.

Meho, L.I. (2006) 'E-mail interviewing in qualitative research: A methodological discussion', *Journal of the American Society for Information Science & Technology*, 57 (10) 1284–95.

Mesch, G.S. & Beker, G. (2010) 'Are norms of disclosure of online and offline personal information associated with the disclosure of personal information online?', *Human Communication Research*, 36 (4) 570–92.

Minichiello, V., Plummer, D. & Macklin, M. (2003) 'Sex in Australia: Older Australians do it too!', *Australian and New Zealand Journal of Public Health*, 27 (4) 466–7.

Minichiello, V., Plummer, D. & Seal, A. (1996) 'The "asexual" older person? Australian evidence', *Venereology: The Interdisciplinary, International Journal of Sexual Health*, 9 (3) 180–8.

NielsenWire (2009) *Six Million More Seniors Using the Web than Five Years Ago*. The Nielsen Company, http://blog.nielsen.com/nielsen/online_mobile/six-million-more-seniors-using-the-web-than-five-years-ago, date accessed 16 December 2009.

Opdenakker, R. (2006) 'Advantages and disadvantages of four interview techniques in qualitative research', *Forum Qualitative Sozialforschung/Forum: Qualitative Social Research*, 7 (4) Art. 11, http://nbn-resolving.de/urn:nbn:de:0114fqs060 4118.

Parks, M. & Floyd, K. (1996) 'Making friends in cyberspace', *Journal of Communication*, 46 (1) 80–97.

Pew Research Center for the People and the Press (1996) Biennial Media Consumption Survey, April 1996. PEW Research Center, Washington, D.C. Available at: http://people-press.org/reports/pdf/127.pdf.

Philbeck, J. (1997) 'Seniors and the Internet', *Cybersociology*, 2 (November 20) http://www.cybersociology. com/files/2_2_philbeck.html, date accessed 7 August 2006.

Richters, J. & Rissel, C. (2005) *Doing it Down Under: The Sexual Lives of Australians* (Sydney: Allen & Unwin).

Seymour, W.S. (2001) 'In the flesh or online? Exploring qualitative research methodologies', *Qualitative Research*, 1 (2) 147–68.

Stieger, S. & Goritz, A.J. (2006) 'Using instant messaging for Internet-based interviews', *Cyberpsychology & Behavior*, 9 (5) 552–9.

Suler, J. (1997) *Psychological Dynamics of Online Synchronous Conversations in Text-Driven Chat Environments*. http://www-usr.rider.edu/~suler/psycyber/text talk.html date accessed 20 March 2007.

Tatano Beck, C. (2005) 'Benefits of participating in Internet interviews: Women helping women', *Qualitative Health Research*, 15 (3) 411–22.

UK Ofcom (2006) *Media Literacy Audit: Report on Media Literacy amongst Older People* (Office of Communications, UK).

Underwood, H. & Findlay, B. (2004) 'Internet relationships and their impact on primary relationships', *Behaviour Change*, 21 (2) 127–40.

Voida, A., Mynatt, E.D., Erickson, T. & Kellogg, W.A. (2004) 'Interviewing over instant messaging' in *Proceedings of the ACM SIGHI Conference on Human Factors in Computing Systems* (CHI '04), Vienna, Austria, pp. 1344–7.

Waite, L.J., Laumann, E.O., Das, A. & Schumm, L.P. (2009) 'Sexuality: Measures of partnerships, practices, attitudes and problems in the national social life, health, and aging study', *Journals of Gerontology Series B: Psychology and Social Sciences*, 64B (Supplement 1): i56–i66.

Weg, R.B. (ed.) (1983) *Sexuality in the Later Years: Roles and Behavior* (New York: Academic Press).

Whitty, M. & Gavin, J. (2001) 'Age/sex/location: Uncovering the social cues in the development of online relationships', *Cyberpsychology & Behaviour*, 4 (5) 623–30.

Wysocki, D.K. (1998) 'Let your fingers do the talking: Sex on an adult chat-line', *Sexualities*, 1 (4) 425–52.

Xie, B. (2005) 'Getting older adults online: The experiences of SeniorNet (USA) and OldKids (China)' in B. Jaegar (ed.) *Young Technologies in Old Hands: An International View on Senior Citizens' Utilization of ICT*, pp. 175–204 (Copenhagen: DJOF Publishing).

Zamaria, C. & Fletcher, F. (2008) *Canada Online! The Internet, Media and Emerging Technologies: Uses, Attitudes, Trends and International Comparisons 2007* (Toronto: Canadian Internet Project).

9
Growing Old for Real: Women, Image and Identity

Mary MacMaster

Introduction

The premise for embarking on my inter-disciplinary research and photographic project, *Growing Old for Real*, was initially an observation that although researchers and theorists had suggested the appearance of older women was judged more harshly than that of ageing men (Sontag, 1972, pp. 29–38; Barrett, 2005, p. 177; Teuscher and Teuscher, 2007, p. 8), there had been little focus on the effect of image on how women age. More recently, interest in this area of study has increased, and my intention was to create not just an illustration, but an integrated context for critique, where inter-disciplinary theory, face-to-face research interviews and a portfolio of creative photography would play an equal part.

The focus was a discourse about ageing which involves the ideas, experiences and opinions of ordinary women. It began with a review of a wide range of literature on questions of image, both visual and the 'pseudo ideals' of advertising images (Boostin, 1977, p. 185), from sociology, gerontology, psychology, psychoanalysis, feminist perspectives and social construction theory. This, along with items from the media, informed themes for a series of face-to-face interviews with ageing women which, in turn, provided issues for my personal and creative response in the form of staged photographic projects. The first project, *Performing Mrs Whistler* grew from a frequent observation from the interview group that their lives and appearances differed greatly from earlier generations, and the second, *Self and Image* stemmed from the ways in which choices about appearance were affected by societal and media pressure.

My aim was to show that photographic images which sit within the domain of art could provide an additional perspective to explore ageing femininity, beyond the more conventional discourses of text and imagery in, for example, advertising, psychoanalysis and gerontology. However, at the outset there were no firm plans for the development of the photographic projects, beyond them comprising a series of staged, conceptual or imaginary images.

Background

Vision informs much of our culture, and is usually regarded as the dominant sense in Western culture and images, expressing as it does, human concerns and interests from the earliest times. Douglas Harper explains that the parts of the brain that process visual information are 'evolutionarily older than parts that process verbal information', and 'evoke deeper elements of human consciousness than do words' (2002, p. 13). In this way, a subtle process of internalisation creates a constantly changing range of ideas and opinions which are then reinforced and filtered by the way society informs new expectations and norms (Jay, 1988, p. 116), which in this case are centred on perceptions and management of the female ageing process. Awareness of the effect of these processes does not guarantee immunity from the effects of media image. As Silvia Kobowlski argues, manipulation and resistance may be understood as fully conscious events:

> It must also be recognised that the unconscious allows manipulation to achieve its effects and so is open to fantasy where fluidity of identification allows the feminine gaze at fashion's image of perfection to become 'a one way prescriptive identification (1990, p. 141).

From the early 1990s a parallel interest in both 'embodiment' and the contribution of image, particularly photography, emerged as issues in understanding processes of ageing (Featherstone & Wernick, 1995; Blaikie & Hepworth, 1997; Blaikie, 1999).

> If the workings of the human imagination are motivated and shaped by culturally determined ideas and practices, then the range of visual images produced [...] provides us with important information concerning the ways in which visual perception of the external appearance of old age [...] are constructed (Blaikie & Hepworth, 1997 p. 102).

Alison Neilson notes in her unpublished 2005 PhD thesis, *Dis/Appearance of the Older Woman,* that photography of ageing femininity is fraught with difficulty. She argues that there is no 'visible alignment' between photographic portraiture and ageing and there can be no positive image of ageing because every image is a reminder of mortality. It is true that the medium of photography is itself closely aligned to the process of ageing for once the image is taken, at whatever stage of life, nothing is ever quite the same again. Everything has 'aged', however imperceptibly, and indeed, has taken another small step towards death.

However, photography has for some time been a popular means, both as a therapeutic approach (Spence & Stanley, 1995) and as a sociological exploration of issues of ageing, and has often taken a documentary line, showing women in real circumstances. Grace Robertson, for example, recorded the ebullience of East End women on a coach outing in the fifties (*Mothers' Pub Outing*, 1954); Graciela Iturbide showed the matriarchal society of a Mexican pueblo (*Juchitan*, 1986), and Gaby Messina made a series of portraits of South American grandmothers in their homes (*Grandes Mujeres*, 2002). Others took a more conceptual line, relying on imagery and association rather than the factual approach of documentary photography. Anne Noggle portrayed the ageing female body as a changing landscape (*Stonehenge Decoded*, 1977) and Mary Kelly's seminal body of indexical work was based on changing female experience (*Interim*, 1984–89). Rosy Martin and Kay Goodridge interrogated assumptions of invisibility in *Outrageous Agers* (2000), and Cindy Sherman produced some grotesque images of herself using prosthetics, wigs and make-up for her *Older Women Series* (2002), an exploration of the pressures placed on ageing women to maintain youthful and beautiful looks.

Although the photography projects for my portfolio are a personal and staged response to issues arising from the interviews there is, inevitably, a tension between the evidential and creative. My intention was to evoke feelings and raise further debate about aspects of female ageing, not with the aim of providing hard and fast 'answers' to the multiple contradictions inscribed within the discourses of ageing femininities, but rather in the hope that the viewer will engage with my portfolio dialectically. Partial conclusions might thus be suggested, but these give rise to further debate. The sociologist, Les Back, notes in *The Art of Listening* that because verbal information alone is no longer adequate to understand an increasingly complex world, 'An imaginative engagement with the social world [is needed], utilising a range of media, verbal and non-verbal forms of representation' (2007, p. 7).

The projects

A series of semi-structured interviews was based on themes arising from theoretical literature and the media. Interviewees were invited to talk about the physical and psychological effects of the menopause, invisibility, gender and generational differences, sexual attraction, feelings about body changes and appearances, and the influence of icons in choosing clothes and make-up. I also asked interviewees for their comments on a number of advertisements which featured ageing media models. At the end of each interview I took a digital photograph of the interviewee and invited immediate reaction, which was usually about the image itself or, occasionally, the process of being photographed.

From the main issues that arose from the interviews, I chose two aspects for the creative photographic projects. The first, generational difference in the process of ageing, became *Performing Mrs Whistler,* which was followed by a participant group discussion about the experience. The second project, *Self and Image,* was based on reaction to personal image and societal pressure to maintain a youthful appearance. Participant feedback in this case was invited by email. Although they were not directly involved in the photographic projects, a few interviewees also took up an invitation to comment via my website www.growingold.co.uk, where all the project images are posted.

The participants

Growing Old for Real involved three separate groups of women – 20 formed the interview group; eight formed the model group for *Performing Mrs Whistler,* and five formed *Self and Image* and they will be considered separately in the following sections under those labels. For ethical reasons I used amateur models rather than women from the interview group. Whilst the inclusion of personal images taken at the end of the interviews would certainly have provided another dimension to the study and the lack might hence be seen as a constraint, I was also concerned about the possible effect of its inhibiting the interviews when highly sensitive and personal information was often discussed. Although happy to talk, several women did not enjoy having their photographs taken and would certainly not have wished them to be seen by others alongside matters usually discussed in private, if at all. For the same reason, amateur models, rather than the interview group, were invited to participate in the photographic portfolio projects.

I used 'snowball sampling' to recruit participants throughout the study. Although, as Atkinson and Flint (2001) point out, snowball sampling can lie on 'the margins of research practice', it also offers 'real benefits in some circumstances' for those who are difficult to access and those 'who require greatest level of trust to be built up' (paragraph 1). Although older women are hardly a hidden section of society, the issue of trust was important because discussion based on the ageing body is both personal and potentially sensitive. Because I, too, am an ageing woman experiencing the effects of living in a highly youth-centred society, mutual contacts were able to assure possible participants of empathy and understanding (Gray et al., 2007).

The interview group

Because the first signs of physiological ageing are frequently noted at about the age of 50 (Woodward, 1999, p. xiii), the lower age limit of my research group was to be 55. Empirical research suggests that women become less concerned about appearance after the age of 75 (Dumas et al., 2005), but because I was unconvinced of this as a generalisation, I chose no upper age limit. Snowball sampling was initiated by asking a friend in her early 60s, who did not wish to be interviewed because she thought I knew her too well, to suggest a possibility. She did so, and also suggested a second who, in turn, was able to find a third, followed by another. There were several links – one interviewee volunteered when I went to talk about my plans at an over-sixties' club, and she passed on two others who agreed to take part. An ex-colleague offered herself and suggested two more names, who found one more each. A fellow student, working on a project with a Caribbean community in another town supplied two names of women who were then able to suggest others. Sometimes the women contacted me directly or gave permission for me to get in touch with them. Two women declined the initial invitation because of ill-health, and one requested a list of themes before the interview because she wished to avoid certain undisclosed areas. Once satisfied that only freely-given information would be used, she agreed to participate.

The interview group, as Table 9.1 shows, covered a broad range of occupations. Some women were still working, occasionally beyond 65, whilst others were unemployed or retired. Some were in relationships, and others were single; some had children and grandchildren, and by their life stage had encountered many different situations. They lived in a range of settings, from urban local authority provision to an apartment in a country mansion. Income and educational experience were equally varied – some had left school at 14 with no formal qualifications, whilst others held vocational training certificates, diplomas or academic degrees. Most were white British, two came from the United States of America, and four were of Caribbean origin although they had lived in England for

Table 9.1   Sample demographics (quotations from this group are denoted by the coded initial and age in bold type)

Code	Age at interview	Ethnic origin	Occupation
S	63	White UK	Artist
P	71	White UK	Retired TV presenter/ex-model
H	65	White UK	Retired factory worker and carer
Bj	83	Black West Indies	Retired factory worker
Gr	88	White UK	Retired cook
K	60	White UK	Secretary
Do	61	White UK	Retired teacher
Br	67	White UK	Cleaner
Qu	65	White UK	Designer and photographer
Ol	62	White UK	Freelance journalist
Ti	60	White UK	Ordained Church of England Clergy
Lm	62	Black West Indies	Retired Civil Servant
We	67	Black West Indies	Retired nurse
Cp	57	White UK	Primary School Head Teacher
Ef	70	White USA	Photographer and writer
U	72	White UK	Volunteer 'green' café
Sn	75	White UK	Retired postal worker
Z	61	White UK	Shop assistant
Js	87	Black West Indies	Retired factory worker
I	68	Mixed race USA	Retired carer
E	65	White UK	School cook

many years. The one thing they all had in common was the experience of female ageing in the 21st century in England.

The process of identifying participants, sending information, completing paperwork for the validating university's ethical requirement, and arranging and carrying out interviews took about a year. As Frida Kerner Furman points out, it is a very 'labour-intensive' process which must be undertaken with careful attention to detail (1997, p. 12). Preliminary information was sent to each prospective participant to outline the role of the interview within the framework of the project as a whole. All comments were to be anonymised; if they wished, participants could see transcripts of their interviews, and also withdraw at any point. They decided where the interview should take place – usually at their homes but three preferred elsewhere. Each participant agreed that the interview, which lasted about an hour, should be recorded on tape. Sections of the interviews were sometimes played back but no changes were made. No one requested transcripts or withdrew, and all signed consent forms for the material to be used.

In-depth, semi-structured, ethnographic interviews were chosen to provide opportunities for both 'a vivid description of life experiences' and a reflexive approach for me in my research as an ageing woman (Westby et al., 2003, pp. 4–5). A wide range of themes for the interviews was developed from literature and items from the media, and often followed a line of discussion initiated by the participant, to allow an active relationship between myself and the interviewee (Hammersley & Atkinson, 2003). They included experience of menopause, 'invisibility', changes in appearance, pressure to maintain a youthful appearance, generational and gender differentials, and responses to advertising images. Some themes received a more robust response than others, which provided a core of consensus for the photographic projects. I began by asking 'When did you first notice signs of ageing?', and while subsequent responses were sought via occasional prompts to provide insight into how and why women approach the ageing process, the women's voices remained very much their own.

Great care was taken to avoid 'leading' the interviewee, however gently, to divulge more than she may wish – this was occasionally a temptation when talking, for instance, about surgical or chemical intervention of appearance, when there may have been some reticence. Most spoke freely and openly, and some were willing to discuss their experiences of surgical intervention and the use of Botox in some detail – one found the opportunity empowering because it was not a usual area for conversation.

The women also commented on a series of prepared images of older female models sourced from magazines, newspapers and advertisements which were shown at the end of the interview. These were chosen to show icons of 'positive ageing' – professional models Twiggy (60), Daphne Selfe (73) and Dove toiletry products 'wrinkled is wonderful' amateur model Irene Sinclair (93). There was also an advertisement for a Mazda car and satnav, showing a deliberately frumpy older woman in plastic rain hat and tweed coat hesitating about giving directions to a motorist.

At the end of each interview I took a digital portrait which was immediately transferred to a laptop computer to enable a response before deletion, once more to protect confidentiality. One woman had initially declined to have her photograph taken but was reassured by being able to witness its removal. This was the only real instance of suspicion I encountered and was based on the experience of a relative of hers whose image had appeared on a poster without his prior knowledge or permission. My interest focused on their *reactions* to their photographs rather than the images themselves and I hoped that this use of photo-elicitation would encourage them to express feelings, whatever they might be, about their respective portraits and appearances (see Chapter 2).

This is what some of the interviewees had to say about their photographs, and relates to the frequently explored issue of the mask of ageing (Rivière, 1929; Sontag, 1972; Featherstone et al., 1991; Woodward, 1999; Tseëlon, 1995; Ballard et al., 2005).

> 'No, I don't like that one. My rosy cheeks, don't like them. It isn't the image I've got of myself...No, that's not me, no...I don't know...I just think it looks a lot older than I feel. It's not the picture I have in my head. I look like a cuddly nanny! I want to be a nanny to my grandchildren but I just don't want that nanny image!' **H. 65.**

> 'I don't like that! I don't look like me...how I perceive me...I think I look fat and I think I see myself as thin! I think it's because I've always been thin and the fat has sort of arrived on my doorstep one day.' **S. 63.**

> 'It doesn't represent me because...I feel 30 or 40 or whatever and I see myself then I see this person who isn't 30 or 40.' **U. 72.**

During the interviews, many of the women said that although ageing women were still compared unfavourably with their male counterparts, they felt that things were considerably better for them than the previous generation. Their expectations were higher and their lifestyles, health care and opportunities different from those of their mothers and grandmothers. As well as often being adversely surprised by their photographic images and/or mirror reflections, they also felt affected by increasing pressure to stave off physical signs of ageing. I chose these issues as a basis for two photographic projects.

The photographic model groups

The first project was based on the oil portrait made by J.M. Whistler of his mother in 1871, and the second used the genre of fashion photography. Both scenarios were created in a photographic studio to which amateur models, with no previous experience of photographic modelling were invited.

The 'snowball' method of recruitment was used once more for the photographic project models. Recruitment for *Performing Mrs Whistler* started with a direct approach to an acquaintance who accepted and passed on the names of three others; an ex-student recommended the project to another and then to a work colleague who suggested others.

The sample included ten women which led to ten portraits to allow a range of responses to the appearance of ageing, but one woman withdrew and another did not turn up at her appointed time. Detailed information was sent in advance to those who accepted. The sample characteristics are summarised in Table 9.2.

Table 9.2 Models for *Performing Mrs Whistler*

Ref	Age at time of shoot	Ethnic Origin	Occupation
F	65	White UK	Textile artist
Cl	63	White UK	Dress designer
Ta	60	White UK	Primary school teacher
Nj	61	White UK	Public relations assistant
Z	68	White UK	Interior designer
T	71	White UK	Retired shop manager
O	61	White UK	Shop assistant
Mj	62	White UK	Charity organiser

A similar procedure was followed for the second project *Self and Image*. This project involved two sessions per model and was conducted a year after *Performing Mrs Whistler*. Once again, there was a last minute withdrawal, and this time, one woman was unable to complete the second half of the project, bringing the total for this project to five – ten images in all. Preliminary information for both photographic projects included reasons for the portraits and their place within the framework of my research study. Consent forms were included, which models signed and returned on the day of each respective shoot, but I also provided extras in case anyone forgot them, which several did.

Table 9.3 Models for *Self and Image*

Ref	Age at time of shoot	Ethnic Origin	Occupation
Fi	66	White UK	Musician
Di	73	White UK	Retired chorus girl and artists' model
Ur	55	White UK	Senior social worker
K	64	Black West Indies	Retired midwife
V	65	White UK	Retired social worker

Issues raised in the interviews concerning social class and appearance were of importance to *Self and Image* which dealt with choices about self-presentation. As Julia Twigg notes, the recognition of an ageing body is 'reflected or resisted, in clothing choices' (2007, p. 290). **I.**, **68**, a retired carer for the elderly, living in social housing, described her choice of clothing like this:

> 'It's just where I can find something that fits! I'm not much of a clothes person really. It's just comfort...just be comfortable.'

Other pragmatic responses, also from interviewees on limited income included:

> 'Fashion don't interest me [sic]...it don't matter to me what anyone else wears – it's what I feel like inside. Always been the same.' **Gr. 88.**

> 'Fashion don't make much difference [sic]. For older women there isn't much fashion...it's what flatters you when we fill out round the waist, fashion goes out the window anyhow. You have to have what fits you.' **H. 65.**

While it would have been interesting to have included similar women in the photographic projects, I received only one such volunteer who withdrew at the last moment, with some apparent relief, because of a work commitment. Two women interviewed said that they did not even look in a mirror and would certainly not relish the prospect of their photographs being looked at by others, a reaction which raises important issues for future research.

Feedback from the models of the photographic projects was obtained in different ways. Most of the *Performing Mrs Whistler* group met shortly after the shoot, but because distances and work schedules prevented that for the *Self and Image* models, comment was invited by email. Three of the five models responded. All models received a print of their portrait which they were able to choose onscreen at the end of their respective sessions.

How the projects developed

Performing Mrs Whistler
During the interviews there had been a strong consensus of opinion about gender and generational difference. Whilst almost all participants believed

that the appearance of older men was still judged more kindly than that of women – and, in fact, did so themselves – they also thought that, compared with their mothers and grandmothers, matters had, overall, changed, albeit with mixed results:

> 'I don't think they [previous generations] had the money and everything like we have now...there weren't any magazines and no television, was there?' **Br. 67**[1]

> 'I think [ageing] was almost more dignified in a way – you grew into your age and it was all accepted – there wasn't this awful thing of trying to look younger, just wasn't on the agenda...no pressures like we have.' **Z. 61**.

Others remarked on the differences in lifestyle – earlier generations had not worked as often as they do now and neither did they socialise in the same way:

> 'My mother never worked. She only worked before she married my father and then all her friends, they never worked. They changed into twinsets and pearls at teatime, four o'clock and it was a totally different sort of life. When my father died she was only fifty and yet she seemed old. Even before he died she seemed much older. She didn't do any of the things we do – partying line, that kind of thing. I mean my mother has never even had a hangover!' **K. 60**.

It was also noted that neither hormone replacement therapy nor surgical or chemical intervention was available for earlier generations; hair products and cosmetics were basic, if used at all and unless they were wealthy, women did not spend much money on themselves. They generally expected to remain in marriages and even though many women did go out to work, some with few resources did not, as one woman from a low-income family remarked:

> 'Years ago our mothers used to stop at home and see after us...father went out to work but mother didn't...she would never dream of going out to work.' **Gr. 88**.

In all, this added up to a recognition of the social construction of ageing femininity, and brought to mind *Arrangement in Black and Grey No. 1, Portrait of the Artist's Mother*, the oil portrait painted by J.M. Whistler of his mother in 1871, which stimulated the development of *Performing Mrs Whistler*.[2] Not only does Whistler's portrait relate closely to the issue of social construction, but it also holds an iconic status within the discourse of visual representation of ageing femininity. Aged 67 at the time of the portrait, considerably younger than the present ages of current icons Joan Collins and Jane Fonda (77 and 76 respectively, at the time of writing), Mrs Whistler would have been considered an old woman because life expectancy at that time was about 50, a good example of the fluidity of the meaning of 'old' (Thane, 2000; Botelho & Thane, 2001; Campbell, 2006). Anna Whistler was to become a model of motherhood and graceful ageing, particularly in the United States. As Kathleen Woodward noted some 128 years later, 'Images shape our view of older women and show women performing age' (Woodward quoting Sobchack, 1999, p. xix), and the Whistler ideal of ageing femininity contributed to expectations of older women well into the 20th century.

A reconstruction of the Whistler portrait scenario was created in the studio and the model group of eight women, all of similar age to Mrs Whistler, was invited to imitate her pose. Each was given a time slot and directions to the studio, where they were met by a colleague who provided general support. Since studios can be intimidating, the women were given time to familiarise themselves with surroundings. With hindsight it may have been helpful to have organised a prior group visit, as experience was to show, when a valuable potential participant withdrew, probably due to a lack of confidence. On the other hand, because none had had any previous photographic modelling experience, they were able to reflect a more personal response than professionals would have provided.

Each portrait took only a few minutes because both camera settings and lighting were the same throughout. Although the studio timetable did not allow the group to meet on the day, they were able to view their portraits – and those taken earlier – immediately, onscreen, and were invited at a later date to discuss the experience of the shoot and receive a print of their chosen portrait. The informal discussion on this occasion centred on the aesthetics of the portraits and how they resembled – or did not – their mothers.

Because it was necessary to recreate a similar ambience in order to highlight differences, the set for *Performing Mrs Whistler* was made with careful attention to position and lighting. Whistler would compose his entire palette before embarking on a painting, making it a matter of some importance for my portraits to be as faithful as possible to the colouring of the original. However, an exact replica of the background

Figure 9.1 Untitled from *Performing Mrs Whistler* 2007. The model chose to wear a jacket that had belonged to her mother

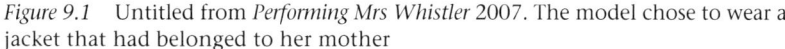

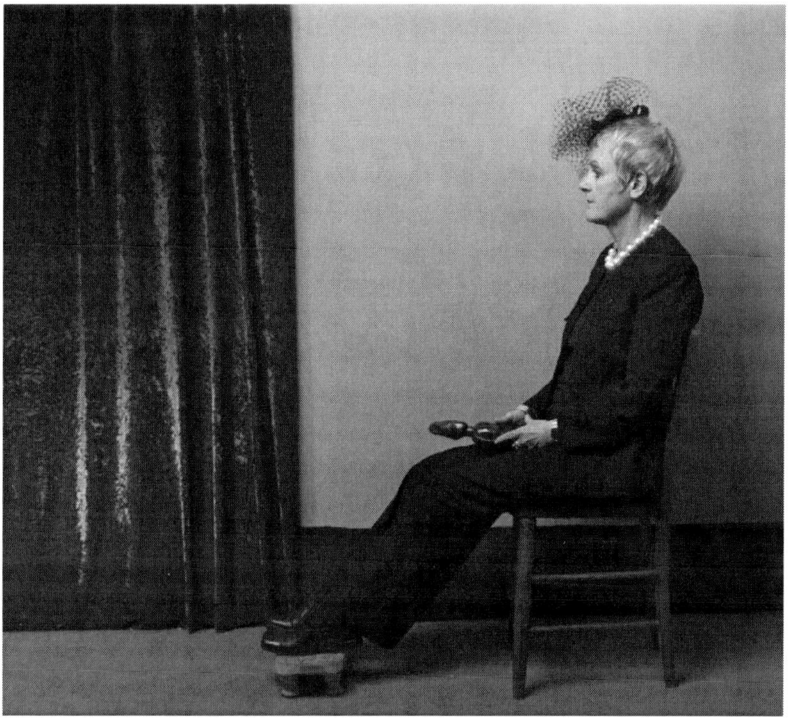

proved impossible, so the eventual choice was a neutral grey. The detail of furniture was less important, as the original Mrs Whistler's skirt covered most of the chair, but because it affected the pose of the sitters and needed to be in keeping with the general scenario, an appropriate piece of furniture was acquired. Similarly the footstool had to be both of the right height and position. A copy of the original portrait initially placed on the back wall was subsequently removed because it distracted the eye from the sitter and provided 'too much information'.

The models were asked to wear black because of the need to restrict the palette – any areas of additional colour would have distanced the Whistler reference. However, black no longer necessarily carries the meaning of mourning as it did in Mrs Whistler's day – indeed it is now something of a fashion statement, often chosen by older women to detract from weight gain. Mrs Whistler is holding a lace handkerchief – perhaps it connotes mourning for her husband who died 22 years previously – so the models

Figure 9.2   Untitled from *Performing Mrs Whistler* 2007. The model is holding photographs of her daughters

were asked to hold an object that was – or had been – important to them. They chose a range of objects from a vibrator to family photographs. Several objects connect with previous or following generations in their families – one woman brought a clock, the only item she has retained from her mother's house; another wore a jacket that had belonged to her mother; there was a rag doll made for a daughter and then a granddaughter. Like Mrs Whistler, the models do not engage with the viewer but, although serious, they appear relaxed and confident. Although their expectations and opportunities are very different, they too are performing their lives, engaging in regular small acts of change that will gradually alter the ways in which older women are perceived both by themselves and others.

Although the differences between my models are clear, the historical comparison with the original lends weight to the question of change over time, because it emphasises the development of similarities and differences. The use of a well-known image is not unusual (e.g. Cindy

Sherman, 1989–90, *The History Portraits*) but some knowledge of the original underpins the interpretation of the viewer without the need for parallel display. The art historian, Griselda Pollock refers to this process as 'reference, deference, difference' (1992, p. 12) and *Performing Mrs Whistler* can be seen, for instance, as a direct response to the original Whistler portrait, whilst the significant differences indicate the originality of my work.

The objects in their hands offer clues to their identities, often, like Mrs Whistler, reflecting family connections – usually between mothers, children, sisters and grandchildren rather than partners and lovers. Mary Kelly notes the significance of female fetish which she suggests may relate to the 'empty nest' syndrome, the need to hold onto connections with children, or perhaps to the body of the mother (1996, p. 74). These connections are implicit but the choice of objects is expressive and lends itself to interpretations such as the vibrator, a direct expression of sexuality, which was substituted in a second shot with a still somewhat phallic skeleton of a horned animal head.[3]

Although the reasons behind the choices are unknown to the viewer, there are strong visual clues to their responses to ageing: some have fashionable, creatively coloured hairstyles, some wear designer fascinators, makeup, red nail varnish, jewellery and stiletto-heeled shoes suggesting a challenge to the traditional stereotypes of ageing; but there is also grey hair and a variety of body size, shape and style – skirts, trousers, dresses. Some of the models were certainly interested in keeping a youthful appearance and whilst others were less concerned, they nonetheless wanted to look their best. Choice of red nail varnish, fishnet stockings and stiletto shoes could be interpreted as a means of retaining sexuality through fetish, but there was no objectification through an overtly sexual pose. This might have been the case had the vibrator been retained, although the shape of the horned animal skeleton could be interpreted (maybe unintentionally) as performing almost the same function, albeit at a much more allusive and less direct level of signification.

When members of the interview group were invited to comment on the photographic projects via my website one woman wrote about *Performing Mrs Whistler*:

'I liked the portraits of the 'mothers'. It showed how we've changed since *our* mothers' days and my gran would have had a fit about the fishnet stockings one of them wore! Only tarts wore those in her day!' **E.65**.

The question of self-presentation was central to *Performing Mrs Whistler* and led to consideration of how women make decisions about appearance, and how they are influenced by societal expectations and media. Some of the women in the interview group had expressed a strong reaction to reflections suddenly caught in a shop window or mirror and to recent photographs, and this led to the development of the second project, *Self and Image*.

Self and image

Many of the women in the interview group recognised Kobowlski's argument about manipulation quoted earlier in this chapter. The following comment was made by a participant when looking at an image of Twiggy modelling clothes for Marks and Spencer:

> 'We all know that's Twiggy and I personally think, oh good on you, you look great. And I know it's a fashion photo and it's touched up so she probably doesn't look like that at all…fair enough…but it is saying to us you too could look like this which of course we couldn't. We never could to be fair; we could never look like her. I think, oh good for M and S, presenting an older woman, but actually they have presented her as a younger woman! But we all know she's older.' **Z. 61.**

Others in the interview group were nevertheless inspired by celebrity images such as those of the fashion model, Twiggy, shown at the interviews and actor, Helen Mirren (whose image was not shown).

> 'It's an aspirational image [Twiggy]…she looks pretty good…'

> I'd like to look like Helen Mirren, dress like her, *be* her…she has a sort of magical style.' **Do. 61.**

> 'There are people I would like to look like – Helen Mirren for example.' **Qu. 65**

The two emergent issues – response to personal portraits and societal pressure on women to maintain a youthful appearance – combined to create the second project, *Self and Image* which featured a second group of five non-professional models aged between 55 and 73 (see Table 9.3). This group included a senior social work officer, a retired social worker, a teacher, a professional musician, and a retired chorus girl who had also worked as an artists' model. Although identification of 'class' is increasingly more fluid, they may well, nonetheless, be seen as predominantly middle-class even though by this stage of their lives they had encountered many different circumstances – for example, the 'chorus girl', originally from a cultured and wealthy background, now at 73, works as a cleaner and carer. Made in two parts, the first half of the project consists of the women in the role of both model and photographer within a studio setting, and was intended to explore freedom to perform as individuals.

Preliminary information sent to the models explained that the intention of the project was to appropriate the genre of fashion photography as an exploration of the significance of personal presentation in the ageing process. Each model was invited to choose her hairstyle, make-up, jewellery and clothes from her own wardrobe. A white background was prepared in the studio, similar to that often used in photographic spreads of fashion magazines. A full-length mirror was placed to the side, out of shot of the camera, which allowed the models firstly to choose the pose, and then to capture the image with a shutter release cable that remains in shot (see Figures 9.3 and 9.4). The viewer thus sees the model looking in the mirror, rather than the reflection itself, a technique used by David Attie in 1977 for his monochrome series of Russian self-portraits. Whilst the fashion photographer usually directs every detail of the pose and

Figure 9.3   Untitled from *Self and Image* 2008

Figure 9.4   Untitled from *Self and Image* 2008

works with a stylist to produce the minutiae of the desired effect, the models on this occasion tried a variety of presentations, and released the shutter when they liked what they saw in the mirror.

Because the work and personal schedules of the models did not allow a post-shoot meeting, they were invited to comment by email – after the first session one wrote:

> 'I found the experience liberating – rather nervous at first but it quickly became fun, reminding me I could still feel excited and alive. I also felt quite empowered and I think the image reflects these feelings. I shall look at it whenever I need to spark myself up.' **Ur. 65.**

The second part of the project was made with the same group of models some months later, when it was possible to use the whole studio. It was to reflect the issue of pressure, because it was directed by me and included an undressed window display mannequin in the background. It was placed there as a reminder of the 'perfect body', a space it often occupies in the minds of many women who consider body-size and shape to be important.

A strip of paper created a faux 'catwalk' down the length of the studio, with a darkened background and studio lights etc. to heighten the impression of 'pressure' of being 'under the lights'. The models were asked to walk along the paper until they reached a marked spot where they rocked gently with one knee bent and one foot behind the other to create an illusion of movement. Hands were hanging loosely by the side and because fashion models usually adopt an impassive expression, they were asked not to smile.

Unlike the first half of the series, the models appeared to engage with the viewer by looking straight into the camera. The set and the genre itself suggested a fashion show of some kind, a display of a 'look' to others. Whilst each woman presented her own individual style, some posed with uncertainty, whilst others appeared to enjoy the opportunity with confidence. Although the direction and pressure of the scenario generally resulted in a far less relaxed presentation than the first group of images, a partially-sighted model (not shown) who was also a musician and artist, found this a much easier way of working, and her portrait was among the most confident. Similarly the ex-chorus girl (Figure 9.3) who had been used to bodily attention throughout her adult life, posed with ease. At first glance it was not immediately obvious that she was 73 because she is tall, blonde, and very slim, with a youthful hairstyle and wearing clothes and shoes that might equally be chosen by younger women wishing to express their sexuality. Moreover it is certainly not obvious that she has had recent knee replacement surgery, an intervention associated with later life.

The model in the second pair of images (Figure 9.4) has chosen a quite different presentation. For the first part of the project, where she had more control over her pose, she chose colourful clothes. She looks relaxed, almost dancing, as though she is really enjoying herself and there is a sense of freedom about the portrait. Dressed more somberly on the 'catwalk' with just a scarf to lighten the effect, she appears much more constrained and tense, as though perhaps the external direction is exerting rather unwelcome pressure.

The first half of *Self and Image* was entirely undirected in terms of clothing and pose and there was no 'retouching' of any of the portraits, apart from a minor crop to exclude the distraction of background objects in the 'catwalk' section of the project. Although the project was not intended to be individually 'therapeutic', the first half can be seen as an exploration of identity, as well as appearance, because the women were more likely to reveal something of themselves. Most were more relaxed and apparently less self-conscious, able to show or conceal as they wished. There was a further opportunity for selection when they were consulted on the choice of image for the photographic portfolio. Again this was influenced by the self each wanted to show, a combination of identity and appearance, affected to a greater or lesser extent by societal approval. In fact all had some interest in clothes and style. All self-presentation is a valid visual statement whether affected or not by societal pressure or, indeed, by the amount of money available to spend on appearance, and so it is scarcely surprising the women were interested in appearance.

Conclusion

The projects are a combination of contributions from the experience of older women and my own interpretations. Although I selected the issues, they were in each case the most robust of the responses within the interviews: generational difference and societal pressure to maintain a youthful appearance, and the interpretation was reinforced by the presentation of two photographic projects. Although asked to appear in a particular scenario, the models made decisions about their self-presentation that reflected their usual lives as well as identity in photographic scenarios.

The accessibility of photography makes a unique contribution to a wide audience. As sociologists have recognised for some time, in a complex world in which information is conveyed less and less by the word (from cinema and advertising, to information technologies and other new media), understanding of the non-verbal takes on a growing importance. Photography plays an essential role in this new universe of meanings, and this is why I have used it to explore contemporary perceptions of the intricacies of later life and ageing in the 21st century. By 'showing, not telling' it is possible to elicit a range of personal avenues of exploration which may sometimes be missed by conventional discourses, and I hope the outcome is a compelling reflection of aspects of ageing femininity which evoke feeling as well as explore theoretical issues.

Notes

1   Initials of interviewees are in bold throughout to differentiate from the photographic models.
2   J.M. Whistler *Arrangement in Grey and Black No 1, Portrait of the Artist's Mother.* Oil on canvas 144 × 162.5 cm. 1871. Musée D'Orsay, Paris.
3   The substitution was made because the model felt uncomfortable about the prospect of the portrait being seen by people she knew. She had brought two items to test her own reactions.

Annotated reading list

Back, L. (2007) *The Art of Listening* (Oxford and New York: Berg).

Back outlines the ways in which sociology is 'a listener's art' and needs 'an imaginative engagement with the social world, utilising a range of media, verbal and non-verbal forms of representation'. He includes a number of projects using a number of methods and resources.

Furman, F.K. (1997) *Facing the Mirror: Older Women and Beauty Shop Culture* (New York and London: Routledge).

Furman uses observation, interviews and photo-elicitation to support her ethnographic study of the effects of ageing experienced by a group of beauty shop clients in the US.

Although published more than ten years ago, the narrative of research methods is informative.

Holland, S. (2004) Alternative Femininities, Body, Age and Identity (Oxford and New York: Berg).

Holland's research considers the effects of ageing on a group of women who define themselves by body modification, tattoos etc. She includes a comprehensive bibliography and a very useful section on research methods.

Martin, R. (2003) 'Challenging invisibility – Outrageous agers' in S. Hogan (ed.) *Gender Issues in Art Therapy* (London and Philadelphia: Jessica Kingsley Publishers).

An exploration of personal ageing through creative photography with Kay Goodridge, which is extended by the development of workshops where groups of older women shared and found ways to express their experiences.

Woodward, K. (ed.) (1999) *Figuring Age: Women, Bodies, Generations* (Bloomington and Indianapolis: University of Indiana Press).

This collection includes the work of Jacqueline Hayden ('Visualising age, performing age') and Anca Cristofovici ('Touching surfaces. Photography, aging and an aesthetics of change'), both reference a number of photographers who have done work on the ageing process.

References

Atkinson, R. & Flint J. (2001) 'Accessing hidden and hard-to-reach populations', *Social Research Update*, 33, http://sru.soc.surrey.ac.uk/SRU33.html.
Attie, D. (1977) *Russian Self-Portraits* (London: Thames and Hudson).
Back, L. (2007) *The Art of Listening* (Oxford and New York: Berg).

Ballard, K., Elston, M. & Gabe, J. (2005) 'Beyond the mask: Women's experiences of public and private ageing during midlife and their use of age-resisting activities', *Health*, 9 (2) 169–87.

Barrett, A. (2005) 'Gendered experiences in mid-life: Implications for age identity', *Journal of Aging Studies*, 19, 163–83.

Blaikie, A. (1999) *Ageing and Popular Culture* (Cambridge: Cambridge University Press).

Blaikie, A. & Hepworth, M. (1997) 'Representations of old age in painting and photography' in A. Jamieson, S. Harper & C. Victor (eds) *Critical Approaches to Ageing and Later Life* (Buckingham and Philadelphia: Open University Press).

Boostin, D. (1977) *The Images* (New York: Atheneum).

Botelho, L. & Thane, P. (eds) (2001) *Women and Ageing in British Society Since 1500* (Harlow: Pearson Educational).

Campbell, E. (2006) *Growing Old in Early Modern Europe* (Aldershot, UK and Burlington, US: Ashgate Publishing).

Dumas, A., Laberge, S. & Straka, S.M. (2005) 'Older women's relations to bodily appearance', *Ageing and Society*, 25 (6) 883–902.

Featherstone, M., Hepworth, M. & Turner, B. (1991) 'Mask of ageing and post-modern life course' in M. Featherstone, M. Hepworth & B. Turner (eds) *The Body* (London: Sage).

Featherstone, M. & Wernick, A. (eds) (1995) *Images of Aging: Cultural Representations of Later Life* (London and New York: Routledge).

Furman, F.K. (1997) *Facing the Mirror: Older Women and Beauty Shop Culture* (New York and London: Routledge).

Gray, P., Williamson, J., Karp, D. & Dalphin, J. (2007) *The Research Imagination; An Introduction to Qualitative and Quantitative Methods* (Cambridge: Cambridge University Press).

Hammersley, M. & Atkinson, P. (2nd edn) (2003) *Ethnography, Principles in Practice* (London and New York: Routledge).

Harper, D. (2002) 'Talking about pictures: A case for photo elicitation', *Visual Studies*, 17 (1) 13–26.

Iturbide, G. (1986) *Juchitan*, www.gruppof.blogspot.com/2008/06.meeting-graciela, date accessed 9 March 2010.

Jay, M. (ed.) (1988) *Vision and Visuality* (Seattle: Bay View Press).

Kobowlski, S. (1990) 'Playing with dolls' in C. Squiers (ed.) *The Critical Image* (London: Lawrence and Wishart).

Neilson, A. (2005) *Dis/Appearance of the Older Woman* (University of York: Unpublished PhD thesis).

Noggle, A. (1977) *Stonehenge Decoded*, www.cddc.vt.edu/host/weidhus/Writing/noggle.htm, date accessed 20 December 2006.

Pollock, G. (1992) *Avant-Garde Gambits 1888–1893* (London: Thames and Hudson).

Rivière, J. (1929) 'Womanliness as masquerade', *The International Journal of Psychoanalysis*, 9, 303–13, reprinted (1991) in A. Hughes (ed.) *The Inner World and Joan Rivière: Collected Papers 1920–1958* (London and New York: Karnac Books).

Sontag, S. (1972) 'The double standard of aging', *Saturday Review of Literature*, 39, 29–38.

Spence, J. & Stanley, J. (1995) *Cultural Sniping: The Art of Transgression* (London: Routledge).

Sherman, C. (2002) *Older Women Series*, www.uklandscape.net/features/Sherman. htm, date accessed 30 January 2008.

Thane, P. (2000) *Old Age in English History: Past Experiences, Present Issues* (Oxford: Oxford University Press).

Tseëlon, E. (1995) *The Masque of Femininity* (London: Sage Publications).

Teuscher, U. & Teuscher, C. (2007) 'Reconsidering the double standard of aging: Effects of gender and sexual orientation on facial attractiveness ratings', *Personal and Individual Differences*, 42 (4) 631–9.

Twigg, J. (2007) 'Clothing, age and the body: A critical review', *Ageing and Society*, 27, 285–305.

Westby, C., Burda, A. & Mehta, Z. (2003) 'Asking the right questions in the right way: Strategies for ethnographical interviewing', *ASHA Leader*, 8.8, www.asha.org/ about/publications/leader-online/archives/2003/q2/f030429b.htm, date accessed 5 May 2011.

Woodward, K. (1999) *Figuring Age: Women, Bodies, Generations* (Bloomington and Indianapolis: University of Indiana Press).

Afterword: Issues, Agendas and Modes of Engagement

Barbara L. Marshall

Several core arguments reverberate through the contributions to this volume. First, 'ageing' and 'later life' are fields in transformation, both as aspects of social life and as domains of academic study. Second, these ongoing transformations have necessitated the development of new theoretical and methodological approaches. Third, despite some notable attempts, a gap continues to exist between studies of ageing and later life, and the 'mainstream' of the social sciences, including discussions of qualitative research. Yet if qualitative research embodies practices that 'make the world visible' (Denzin & Lincoln, 2011a), then surely neglect of the social relations of ageing and the domain of later life impoverishes this enterprise.[1]

As Miranda Leontowitsch outlines in her introduction, political economy and biomedical approaches have dominated research on later life, focusing on measurement and quantification. As a number of contributors note, the predominance of these perspectives has not just shaped methodological preferences, but has framed issues related to later life fairly narrowly as social policy and health concerns. While not denying that these are significant, contributors to this volume draw our attention to a more expansive social and cultural terrain that demands exploration.

Taken together, the chapters collected here are a good indication of the vibrancy and interdisciplinary character of qualitative research into later life. The contributors come from different disciplinary locations (including sociology, social work, health studies and fine arts), and ground their insights in extensive research practice, collectively drawing on years of experience over a wide range of contexts. As such, they are able to unpack the complexity of research with 'older' populations and on the terrain of 'ageing' in contemporary Western societies. That the very terms 'older' and 'ageing' are problematised throughout the volume underscores that more than methodological technique is up for discussion here. While a wide range of practical methodological issues *are* discussed – including the negotiation of access to marginalised or hidden populations, recruitment of participants, gaining informed consent, and managing the ethical and emotional aspects of research on sensitive subjects – I want to focus in this brief Afterword on the importance of the contributions to this volume for opening up questions that have the potential to nurture new theoretical insights that might in turn inform future research initiatives.

In reflecting on the research strategies and issues discussed here, I am reminded of Gareth Morgan's conceptualisation of research as 'modes of engagement'. As he put it 'A view of research as engagement stresses that research is not just a question of methodology...but part of a wider process that constitutes and renders a subject amenable to study.' It also 'emphasizes the importance of understanding the network of assumptions and practices that link the researcher to the phenomenon being investigated' (1983, p. 19). Such a perspective is helpful in understanding that methodological questions are never just about method. Three themes in particular seem to illustrate this conceptualisation of 'research as engagement' – problematising 'later life' as a field of study, a sustained focus on embodiment, and the ways in which these first two themes necessitate creativity and diversity in choice of research strategies.

Problematising 'later life'

Contributors to this volume have unsettled the notions of 'ageing' and 'old age' in a number of ways. As Paul Higgs recounts in Chapter 1, broad economic, social and cultural shifts have remapped the terrain of 'later life', rendering the distinction between adulthood and 'old age' problematic. Life courses, no longer divided by the conventional markers of childhood, adulthood and old age, are now viewed as more individualised and open to reconstruction through choices about work, leisure and consumption. Most chapters highlight diversity and heterogeneity in later life, illustrating the extent to which age intersects with other axes of social difference (such as gender). Some focus on groups now more likely to experience 'later life' due to increased longevity and biomedical advances (e.g. Chapters 3 and 5), as well as those related to global patterns of migration resulting in more ethnically diverse societies (see e.g. Chapter 4). Significant shifts are occurring and considerable diversity exists, then, in who might be considered 'old' and when, and what cultural resources are available for making sense of experiences of ageing and the construction of ageing identities.

The theme of problematising later life also extends to the insistence of contributors that we expose and critically engage with the 'domain assumptions' (Gouldner, 1971) of research on later life. As Gouldner argued, there are both technical and 'extra-technical' aspects to social research and theorising – it is the latter that encompass the dimensions of situation, experience and biography of the researcher. In other words, we need to interrogate not only the social world, but the tacit assumptions that frame our inquiry. As several contributors note, reflection on assumptions about ageing – sometimes manifesting as internalised age*ism* – was pivotal in extending their inquiry to previously neglected issues (e.g. contributions by Hurd Clarke, Lowton, Leontowitsch, Bigby, and Malta).

A focus on embodiment

The second theme that crosscuts the volume – the focus on embodiment – is one which invokes a constellation of questions about method, theory and substantive research agendas. Conceptualising embodiment in later life has sometimes been limited by biopolitical concerns with the management of ageing bodies. The more expansive research agendas represented in this volume understand ageing bodies as gendered, sexual, racialised, of differing health statuses and abilities, potentially creative and sometimes even pondering the end of their bodily lives.

In her chapter devoted to this topic, Hurd Clarke crystallises a number of core issues related to embodiment that are picked up by subsequent chapters. First, noting the relatively recent sustained attention to embodiment in research with older people, she identifies some of the key areas where qualitative research has contributed to our understanding of embodied experiences of late life – including health, illness, appearance and the negotiation of the bodily aspects of everyday life – and suggests important ways in which these need to be more fully elaborated. She also goes beyond a stock-taking of research insights and potential agendas to engage in a critically reflexive consideration of embodiment, not just as a topic of research, but as a potential site of internalised ageism and emotional sensitivity for both research participants and researchers. Her candid reflections on her awareness of her own embodiment in relation to that of her participants are rich with potential for understanding research encounters as social relationships. This insight is extended in other chapters as the authors explore research practice as both embodied and performative, for all those concerned.

For example, Zubair, Martin and Victor draw attention to the embodied nature of research relations as they provide a detailed reflection on Maria Zubair's active negotiation of a gendered, Pakistani identity in her research with older Pakistani men and women. The many challenges faced in her engagement with this under-researched group (low levels of literacy, access challenges related to gendered spatial organisation, lack of familiarity with and/or distrust of academia and 'white officialdom') underscored the extent to which access and negotiation was an ongoing process, centrally hinging on her visible, embodied identity.

Other contributors echo this reflexivity about the embodied nature of research-as-engagement, for example, Leontowitsch, who as a young woman reflects on the complexity of exploring issues of health and leisure with older men, or Lowton, who worries about the physical isolation and potentially emotional distress that may be experienced by those transcribing interviews with participants they have not met.

Diversity and creativity in modes of engagement

The emphasis on change and diversity in later life, and a reflexive approach to embodiment as a core issue for research agendas and modes of engagement has, of necessity, demanded that researchers develop a suite of diverse, imaginative and multifaceted research strategies.[2] Some of this innovation may be prompted by practical concerns with accommodating differing abilities of respondents to read, hear or understand conventional written or spoken interactions. Seymour, for example, discussing the use of focus groups for research with sensitive topics like end-of-life care, advocates a 'combination of written, verbal and pictorial information' (p. 134) be incorporated into research design. Other innovations are aimed at providing choice and/or privacy to respondents in how they choose to participate. Malta, for example, provided respondents a choice of face-to-face, telephone or online interviews in researching their experience with romantic and intimate relationships. Lowton, researching the transition of young adults with cystic fibrosis to adult and end of life care, advocates multiple methods, including face-to-face interviews, focus groups, letters and self-reported interviews to develop a richer understanding of a population that 'may not wish to be considered old, but…are positioned within the same processes and dynamics as those who are' (p. 53). Others, such as MacMaster, deploy imaginative and creative methods to generate new ways of understanding the complex meanings associated with later life in ways that link the personal and the political. In one strand of her project, she offers participants the opportunity to express visually their understandings of their own experience and identity as women in contrast to their mothers, tapping into the connections between their biographies and their more immediate and embodied experience of social and cultural change. In another strand, participants visually represent shifts in their identity as interpellated through the gaze of mainstream media.

All of these examples demonstrate that participants in the research projects reviewed here are not just sources of data – they are generative sources of understanding regarding questions that cut to the heart of understanding social worlds. It is to the challenges that researching later life poses for some of these larger questions that I now turn.

Continuing challenges and new agendas

If there is one *leitmotif* that characterises this volume, it is the unsettling of conventional assumptions about 'age' and 'ageing' – whether through critical analysis of the socio-cultural context of ageing, of the shifting composition of ageing populations, or of assumptions about ageing. This underscores the importance of qualitative research for posing and exploring new questions and expanding agendas for research.

Across the variety of research sites and populations discussed, all of the researchers represented in this volume are united by a critical interest in age *relations*. That is, their engagement is rooted in a recognition of age as a social organising principle, and aimed at identifying and analysing sites of the social production and reproduction of age-related differences and inequalities (Calasanti, 2003). All are interested in understanding how these processes are undergoing change in contemporary Western societies, how they are connected to other relations of difference and inequality, and how these interconnections continue to be reconfigured. Acknowledging the diversity of ageing populations means greater attention to the production and negotiation of difference in research, and these issues resonate throughout this collection. The legacy of feminist methodological debates is apparent here – whether explicit, as it is in several chapters, or more diffusely through considerations of situatedness, power and accountability in research relationships. It is a truism that age, as a social relation, is always constructed and experienced in and through other social relations such as gender, race, (dis)ability (and vice versa), but integrating such intersectional insights remains a challenge. A complex matrix of social relationships, including age, is now recognised as socially constructed and regulated across a range of institutions and contexts, but there is a need for more research that brings these insights together. This is certainly not an original insight, but one which deserves reiteration. This is particularly the case given the extent to which the cultural terrain on which ageing occurs continues to shift and generate new questions.[3]

A focus on the social construction of age relations as these have been reconfigured in late modern consumer societies is nicely summarised by Gilleard and Higgs' phrase '...the liberation of "ageing" from "old age"' (Gilleard & Higgs, 2005, p. 161). As Higgs suggests in his chapter, this opens up questions of diversity, individualisation, choice and agency that have not traditionally been associated with later life. These questions frame a number of problematic issues that demand further exploration. For example:

- 'Individualisation' as a hallmark of contemporary social life, may be complicated for older people by political and bureaucratic interest in the management of ageing populations, and this may be *de*individualising. Despite the acknowledged destandardisation and fluidity of lifecourses, chronological age continues to justify the allocation of social resources and entitlements and to frame research populations and problems.[4] Lowton's and Bigby's contributions to this volume clearly demonstrate the inadequacy of chronological age in classifying 'older' persons or those in 'late life' in terms of their need for services. From another perspective, Amanda Grenier (2007), suggests that one of the greatest difficulties for researchers is 'to address and account for the gap between the existing age and generational classifications used in practice and the various ways older people interpret, define and experience later life' (p. 722).
- Because cultural discourses of the 'third age' and 'positive ageing' emphasise choice, individualisation and agency, they become problematic for those whose capacity to construct their own late-life biographies is constrained – this would include many constituencies discussed in this volume. While opportunities for choice and agency are certainly not distributed evenly, there is much here to suggest that we might think more creatively about how to conceptualise agency in later life (see also Tulle, 2004). For example, qualitative research can open our eyes to agency where we might least expect it – as in Bigby's research on those with intellectual disabilities 'whose horizons had broadened after the death of their parent in ways that could never have been envisaged' (p. 96).
- Destabilising the distinctions between adulthood, middle age, old age, late life and so on raises a myriad of questions about the boundary work required to sustain positive ageing identities and how these vary by social location. What Higgs refers to in his chapter as the active and

agency-filled cultural terrain of the 'third age' is in large part defined in relation to the spectre of a 'fourth age' of infirmity and lack of agency (see also Gilleard & Higgs, 2010, 2011). The distinction between the active and agency-filled third age and a fourth age of decline and decrepitude is qualitative, embodied, and based on an expanding array of functions, capabilities, appearances and strategies of risk-management. The ways in which different groups negotiate, take up, refuse or transform the available discursive and technological resources for engaging in this boundary work awaits much further research.

Conclusions

As the contributions to this volume demonstrate, both the composition of those deemed 'old' and the contexts for living in 'later life' have changed, and continue to change. The result is a host of new questions to explore, new methodologies for exploring them, and a need for reflection on the process, products and uses of such research. I have suggested that qualitative research has an important role to play in deepening the analysis of age relations as social relations, and rethinking questions of agency in later life. These are interlinked and continually unfolding agendas that have the potential to make important contributions to social theory and critical analysis of contemporary life more generally.

There is a hopeful thread here as well. As Higgs suggests in Chapter 1, 'there is a richness to older peoples' lives that not only repays attention, but also tells us much more about the complexity of modern societies' (p. 18). Against the contradictions of, on the one hand, celebratory representations of active, sexy and self-sufficient 'third-agers' as consumers and, on the other hand, alarmist demographic and policy treatises predicting drained resources and thinly-stretched social safety nets, we are encouraged to pause and 'exercise our imaginations' (Bigby, p. 84) and consider the generative possibilities of understanding the complexity and diversity of late life experiences.

Notes

1   The massive *Sage Handbook of Qualitative Research* (Denzin & Lincoln, 2011b), includes discussions of feminism, queer theory, transnational relations, Asian epistemologies and disability communities as these related to the conceptualisation and practice of qualitative research, but not one of the 43 chapters focuses on age or intergenerational relations.

2   Other recent and instructive examples of creative research strategies in exploring later life include Linn Sandberg's (2011) deployment of 'body diaries' in her research on ageing men and sexuality; Wendy Martin's (2012) use of photo-

graphy to explore perceptions of health and risk; Naomi Richards, Lorna Warren and Merryn Gott's (2011) participatory research aimed at creating 'alternative images' of ageing women; and Meika Loe's (2011) multi-textured ethnography of elders (85 years and older) living in the community.
3 For my own agenda in taking up some of these questions as they relate to ageing and sexuality, see Marshall (2011).
4 Stephen Katz (2010) has usefully reviewed the history of chronological and functional definitions of age.

References

Calasanti, T. (2003) 'Theorizing age relations' in S. Biggs, A. Lowenstein & J. Hendricks (eds) *The Need for Theory: Critical Approaches to Social Gerontology*, pp. 199–218 (New York: Baywood Press).

Denzin, N.K. & Lincoln, Y.S. (2011a) 'Introduction: The discipline and practice of qualitative research' in N.K. Denzin & Y.S. Lincoln (eds) *The Sage Handbook of Qualitative Research*, pp. 1–20 (London: Sage).

Denzin, N.K. & Lincoln, Y.S. (eds) (2011b) *The Sage Handbook of Qualitative Research (Fourth Edition)* (London: Sage).

Gilleard, C. & Higgs, P. (2005) *Contexts of Ageing: Class, Cohort and Community* (Cambridge: Polity).

Gilleard, C. & Higgs, P. (2010) 'Ageing without agency: Theorizing the fourth age', *Ageing & Mental Health*, 14 (2) 121–8.

Gilleard, C. & Higgs, P. (2011) 'Ageing abjection and embodiment in the fourth age', *Journal of Ageing Studies*, 25 (2) 135–42.

Gouldner, A. (1971) *The Coming Crisis in Western Sociology* (New York: Basic Books).

Grenier, A. (2007) 'Crossing age and generational boundaries: Exploring inter-generational research encounters', *Journal of Social Issues*, 63 (4) 713–27.

Katz, S. (2010) *Measuring Age Beyond Chronological Age: Functions, Boundaries, Reflexivity*. Paper presented at the Gerontology Society of America 63rd Annual Scientific Meeting, New Orleans.

Loe, M. (2011) *Ageing Our Way: Lessons for Living from 85 and Beyond* (New York: Oxford University Press).

Marshall, B.L. (2011) 'The graying of "sexual health": A critical research agenda', *Canadian Review of Sociology*, 48 (4) 390–413.

Martin, W. (2012) 'Visualizing risk: Health, gender and the ageing body', *Critical Social Policy*, 32 (1) 51–68. doi: 10.1177/0261018311425980.

Morgan, G. (1983) 'Research strategies: Modes of engagement' in G. Morgan (ed.), *Beyond Method: Strategies for Social Research* (London: Sage).

Richards, N., Warren, L. & Gott, M. (2011) 'The challenge of creating "alternative" images of ageing: Lessons from a project with older women', *Journal of Ageing Studies*, 26, 65–78.

Sandberg, L. (2011) *Getting Intimate: A Feminist Analysis of Old Age, Masculinity and Sexuality* (Linkoping, Sweden: Linkoping University).

Tulle, E. (2004) 'Rethinking agency in later life' in E. Tulle (ed.) *Old Age and Agency*, pp. 175–89 (New York: Nova Science Publishers).

Index

Locators shown in *italics* refer to figures and tables.

204